Conceptual Guide
General Chemistry

Seventh Edition

Darrell D. Ebbing
Wayne State University

Steven D. Gammon
University of Idaho

Houghton Mifflin Company   Boston   New York

Editor in Chief: Kathi Prancan
Senior Sponsoring Editor: Richard Stratton
Senior Development Editor: Marianne Stepanian
Manufacturing Manager: Florence Cadran
Marketing Manager: Andy Fisher
Editorial Assistant: Marisa R. Papile

Printed in the U.S.A.

ISBN: 0-618-11839X

123456789-BB - 05 04 03 02 01

Contents

To the Instructor

With the sixth edition of General Chemistry, we added Concept Checks within the text and a section entitled Conceptual Problems to the end-of-chapter problems. In the current, seventh edition, you will find that the program of Concept Checks and Conceptual Problems has been expanded and improved. In addition to performing a revision and review of the problems from the sixth edition, many new problems have been introduced that reflect the increased emphasis on molecular level thinking in the text.

We developed this guide to accompany the Concept Check and Conceptual Problem features of the text. The guide gives complete solutions to all of these problems and adds a "Concept Target," which essentially summarizes the concept or concepts underscored in the problem.

We should explain what we mean by "conceptual problems" and why we think they have a purpose in the general chemistry course. Problem solving has long been an integral part of the general chemistry course. In part, this stems from the nature of chemistry as a science. Chemists are problem solvers; so if you want to talk about what chemists do, you need to underscore this problem-solving aspect. But instructors cite other reasons, for example, the need to develop students' basic skills (writing chemical formulas, balancing chemical equations, and so forth). Another reason cited for adding problem solving to the general chemistry course is to give the student some appreciation for the quantitative nature of much of present-day chemistry.

Although most instructors would agree that problem solving is important, some have begun to criticize what they feel is a trend toward overemphasizing the numerical aspects of problem solving while underemphasizing the chemical concepts that underlie these problems. Whether they agree with this or not, most instructors would probably agree that students do themselves a disservice by trying to memorize a series of steps for each of the different kinds of problems they think they might encounter. Instructors would much prefer that students approach problem solving by first asking themselves, what is the chemistry that is involved in this problem? Our question then is, how do we teach students to approach problem solving by "thinking chemistry" rather than by using some rote process. We think that adding conceptual problems to the mix of assigned problems will help instructors teach proper problem-solving skills.

By a conceptual problem, we mean a problem so stated that it forces the student to think about the chemical concepts involved. As we noted in the Preface to the text, we have phrased questions to force a thoughtful answer, sometimes by adding a slightly different spin to the question and by not asking for a numerical answer that might lead the student to look for an algorithm, or memorized series of problem-solving steps.

Consider the following conceptual problem.

Three 3.0-L flasks, each at a pressure of 8.78 mmHg, are in a room. The flasks contain He, Ar, and Xe, respectively. Which of the flasks contains the most atoms of gas?

A student who has memorized how to solve various ideal-gas-law problems might try to calculate the moles, then the number of molecules, of gas. He or she would immediately run into a problem, of course, because no temperature is given. But no calculation is needed. Since the flasks are presumably at the same temperature, Avogadro's law would say that each flask contains the same number of molecules. It is our hope that by involving students in such conceptual problems, they would begin to approach all problem solving by thinking first about the concepts involved and then planning a problem solution, rather than blindly trying to work through a solution by rote.

These conceptual problems, therefore, differ from the traditional numerical ones in looking for qualitative rather than quantitative solutions. Many real-world problems are of this sort. We hasten to add that traditional quantitative problems continue to play an essential role in chemistry instruction. It is our intent that these conceptual problems will add a needed balance to the problem mix and will help instructors emphasize the value of a strong conceptual foundation.

To the Student

To emphasize the importance of conceptual thinking in problem solving, the authors of your text developed Concept Checks and Conceptual Problems. These questions focus on having you apply conceptual information to solve problems. These problems will often have you visualize concepts, draw pictures, write out answers, or provide reasonable numerical estimates without performing calculations. In solving these problems, you will develop a balanced understanding of chemistry that includes concepts with problem-solving skills. These problems will introduce you to a variety of situations in which you might encounter chemistry and at the same time hone your skills in critical thinking. Finally, we think that the skills you learn in solving these conceptual problems will be just the skills you need to solve the numerical problems that you will also encounter in the course.

How to Use this Guide

This *Conceptual Guide* is structured to present you, chapter by chapter, with all of the Concept Checks and Conceptual Problems from the text along with a "Concept Target" and in-depth answers. An example of the problem layout for the *Conceptual Guide* is shown below.

Concept Check 1.1

Matter can be represented as being composed of individual units. For example, the smallest individual unit of matter can be represented as a single circle, •, and chemical combinations of these units of matter as connected circles, ••, with each element represented by a different color. Using this model, place the appropriate label – element, compound, or mixture – on each container.

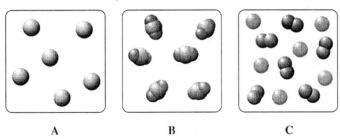

A   B   C

Concept Target

- Begin the process of atom (microscopic) visualization.
- Make a link between a written description and a model.

Solution

The box on the left contains a collection of identical units; therefore, it must represent an element. The center box contains a compound since a compound is the chemical combination of two or more elements (three elements in this case). The box on the right contains a mixture since it is made up of two different substances.

One of the keys to solving any kind of problem, be it numerical or conceptual, is for you to first identify why the question is being asked and what you need to know to answer the question. The Concept Targets associated with each of the problems in this guide are designed to complete the sentence: "The purpose of this problem is to…." When you have identified the Concept Target you probably will know the information that you need to assemble to answer the question. Identifying the Concept Target is a skill that has broad application and is one that you can apply when reading a passage in the text, working a sample exercise, etc.

Here are some steps that you can take to get the most out of using this guide when solving the Concept Checks and Conceptual Problems.

1. When you prepare to answer the Concept Check or Conceptual Problem from the text, try to identify the concept(s) that the problem is targeting. As mentioned above, read the problem and try to complete the sentence: The purpose of this problem is to…. Once you have identified the Concept Target of the problem, then proceed to try and answer the question using the appropriate concepts and skills.

2. If you have difficulty in identifying the Concept Target for the problem, refer to this guide. Hopefully, the concept target will provide enough information to allow you to attack the problem or, at the least, dig out what you need from the text or your notes to arrive at a solution.

3. Once you have come up with a solution, or if you are stuck on finding the answer, refer to the guide for a detailed solution.

Because concept-based problems are a new and different experience, you might find them to be quite difficult at first; initially, you may find yourself stuck on step 1 above. However, as you progress through the material and continue to work with the Concept Checks and Conceptual Problems, you will likely be able to answer most of the concept-based questions without much help from the guide.

We hope that working through the Concept Checks and Conceptual Problems leads you to a greater understanding and appreciation of the fascinating field of chemistry. We also are certain that the approaches and critical thinking skills developed in solving all types of chemistry problems will serve you well in your future endeavors.

Chapter 1

Chemistry and Measurement

Concept Check 1.1

Matter can be represented as being composed of individual units. For example, the smallest individual unit of matter can be represented as a single circle, •, and chemical combinations of these units of matter as connected circles, •• , with each element represented by a different color. Using this model, place the appropriate label – element, compound, or mixture – on each container.

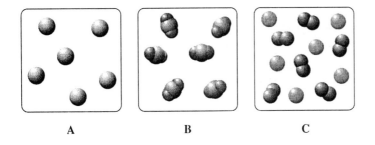

A   B   C

Concept Target

- Begin the process of atom (microscopic) visualization.
- Make a link between a written description and a model.

Solution

The box on the left contains a collection of identical units; therefore, it must represent an element. The center box contains a compound since a compound is the chemical combination of two or more elements (three elements in this case). The box on the right contains a mixture since it is made up of two different substances.

Concept Check 1.2

a. When you report your weight to someone, how many significant figures do you typically use?
b. What is your weight with two significant figures?
c. Indicate your weight and the number of significant figures you would obtain if you weighed yourself on a truck scale that can measure in 50-kg or 100-lb increments.

Concept Target

• Identify rules governing significant figures and the situations where these rules apply.
• Illustrate that significant figures are not really a "foreign" idea; most people have encountered and used significant figures before.

Solution

a. For a person who weighs under 100 pounds, two significant figures are typically used, although one significant figure is possible (for example, 60 pounds). For a person who weighs 100 pounds or more, three significant figures are typically used to report the weight (given to the whole pound), although people often round to the nearest unit of 10, which may result in reporting the weight with two significant figures (for example, 170 pounds).
b. 165 pounds rounded to two significant figures would be reported as 1.7×10^2 pounds.
c. For example, 165 pounds weighed on a scale that can measure in 100-lb increments would be 200 pounds. Using the conversion factor 1 lb = 0.4536 kg, 165 pounds is equivalent to 74.8 kg. Thus, on a scale that can measure in 50-kg increments, 165 pounds would be 50 kg.

Concept Check 1.3

a. Estimate and express the length of your leg in an appropriate metric unit.
b. What would be a reasonable height, in meters, of a three-story building?
c. How would you be feeling if your body temperature was 39°C?
d. Would you be comfortable sitting in a room at 23°C in a short-sleeve shirt?

Concept Target

• Express measurements of familiar quantities using the metric system.

Solution

a. If your leg is approximately 32 inches long, this would be equivalent to 0.81 m, 8.1 dm, or 81 cm.
b. One story is approximately 10 feet, so three stories is 30 feet. This would be equivalent to approximately 9 m.
c. Normal body temperature is 98.6°F, or 37.2°C. Thus, if your body temperature is 39°C (102°F), you would feel like you have a moderate fever.
d. Room temperature is approximately 72°F, or 22°C. Thus, if you were sitting in a room at 23°C (73°F), you would be comfortable in a short-sleeve shirt.

Concept Check 1.4

You are working in the office of a precious metal buyer. A miner brings you a nugget of metal that he claims is gold. You suspect that the metal is a form of "fool's gold" called marcasite, which is composed of iron and sulfur. In the back of your office, you have a chunk of pure gold. What simple experiments could you perform to decide whether or not the miner's nugget is gold?

Concept Target

• Realize that some physical properties can be relied on to determine the identity of a pure substance.

Solution

Gold is a very unreactive substance, so comparing physical properties is probably your best option. However, color is a physical property that you cannot rely on in this case to get your answer.

One experiment that you could perform is to determine the densities of the metal and the chunk of gold. You could measure the mass of the nugget on a balance and the volume of the nugget by water displacement. Using this information, you could calculate the density of the nugget. Repeat the experiment and calculations for the sample of gold. If the nugget is gold, the two densities should be equal and be 19.3 g/cm^3.

Also, you could determine the melting points of the metal and of the chunk of pure gold. The two melting points should be the same (1338K) if the metal is gold.

Conceptual Problem 1.19

a. Sodium metal is partially melted. What are the two phases present?
b. A sample of sand is composed of granules of quartz (silicon dioxide) and seashells (calcium carbonate). The sand is mixed with water. What phases are present?

Concept Target

• Differentiate the phases of matter.

Solution

a. Two phases: liquid and solid.
b. Three phases: liquid water, solid quartz, and solid seashells.

Conceptual Problem 1.20

A material is believed to be a compound. Suppose you have several samples of the material obtained from various places around the world. Comment on what you would expect to find upon observing the melting point and color for each sample. What would you expect to find upon determining the elemental composition for each sample?

Concept Target

• Compounds, regardless of origin, have constant composition and identical physical and chemical properties.

Solution

If the material is a pure compound, all samples should have the same melting point, the same color, and the same elemental composition. If it is a mixture, there should be a difference in these properties depending on the composition.

Conceptual Problem 1.21

You need a thermometer that is accurate to $\pm 5°C$ to conduct some experiments in the temperature range of 0-100°C. You find one in your lab drawer that has lost its markings.
a. What experiments could you do to make sure your thermometer is suitable for your experiments?
b. Assuming that the thermometer works, what procedure could you follow to put a scale on your thermometer with the desired accuracy?

Concept Target

- Emphasize that when temperature scales are created they are based on defined temperature points.
- Illustrate that measuring devices (a thermometer in this case) can be calibrated based on these defined points.

Solution

a. You need to establish two points on the thermometer with known (defined) temperatures, for example, the freezing point (0°C) and boiling point (100°C) of water. You could first immerse the thermometer in an ice-water bath and mark the level at this point as 0°C. Then, immerse the thermometer in boiling water and mark the level at this point as 100°C. As long as the two points are far enough apart to obtain readings of the desired accuracy, the thermometer can be used in the experiments.

b. You could make 19 evenly spaced marks on the thermometer between the two original points, each representing a difference of 5°C. You may divide the space between the two original points into fewer spaces as long as you can read the thermometer to obtain the desired accuracy.

Conceptual Problem 1.22

You are teaching a class of second graders some chemistry, and they are confused about an object's mass versus its density. Keeping in mind that second graders don't understand fractions, how would you explain these two ideas and illustrate how they differ? What examples would you use as a part of your explanation?

Concept Target

- Demonstrate a complete understanding of density by developing an explanation without the aid of mathematics.

Solution

The mass of something (how heavy it is) depends on how much of the item, material, substance, or collection of things you have. The density of something is the mass of a specific amount (volume) of an item, material, substance, or collection of things. You could use 1 kg of feathers and 1 kg of water to illustrate that they have the same mass yet have very different volumes; therefore, they have different densities.

Conceptual Problem 1.23

Say you live in a climate where the temperature ranges from -100°F to 20°F and want to define a new temperature scale, YS (YS is the "Your Scale" temperature scale), that defines this range as 0.0 to 100.0°YS.
a. Devise an equation that would allow you to convert between °F and °YS.
b. Using your equation, what would be the temperature in °F if it were 66°YS?

Concept Target

- Develop an equation to relate a known temperature scale with a new temperature scale using the approach presented in section 1.6 of the text.

Solution

a. In order to answer this question, you need to develop an equation that converts between °F and °YS. To do so, you need to recognize that one degree on the Your Scale does not correspond to one degree on the Fahrenheit scale, and that -100°F corresponds to 0° on Your Scale (different "zero" points). As stated in the problem, in the desired range of 100 Your Scale degrees, there are 120 Fahrenheit degrees. Therefore, the relationship can be expressed as 120°F = 100°YS, since it covers the same temperature range. Now you need to "scale" the two systems so that they correctly convert from one scale to the other. You could set up an equation with the known data points and then employ the information from the relationship above. For example, to construct the conversion between °YS and °F, you could perform the following steps.

Step 1:   °F = °YS
Not a true statement but one that you would like to make true.

Step 2: $°F = °YS \times \dfrac{120° F}{100° YS}$

This equation takes into account the difference in the size of a temperature unit on each scale but will not give you the correct answer because it doesn't take into account the different zero points.

Step 3:   By subtracting 100°YS from your equation from Step 2, you now have the complete equation that converts between °F and °YS.

$$°F = (°YS \times \dfrac{120° F}{100° YS}) - 100°F$$

b. Using the relationship from part a., 66°YS is equivalent to

$$(66°YS \times \dfrac{120° F}{100° YS}) - 100°F = -20.8°F = -21°F$$

Conceptual Problem 1.24

You are presented with a piece of metal in a jar. It is your job to determine what the metal is. What are some physical properties that you could measure in order to determine the type of metal? You suspect that the metal might be sodium. What are some chemical properties that you could investigate (see section 1.4 for some ideas)?

Concept Target

• Ensure a firm grasp of the concepts associated with physical and chemical properties.

Solution

Some physical properties that you could measure are density, hardness, color, and conductivity. Chemical properties of sodium would include reaction with air, reaction with water, reaction with chlorine, reaction with acids, bases, etc.

Conceptual Problem 1.25

You have two identical boxes with interior dimensions of 8.0 cm x 8.0 cm x 8.0 cm. You completely fill one of the boxes with wooden spheres that are 1.6 cm in diameter. The other box gets filled with wooden cubes that are 1.6 cm on each edge. After putting the lids on the filled boxes, you then measure the density of each. Which one is more dense?

Concept Target

• Explore the concept of density.
• Develop a mental picture of the boxes and their contents in order to solve the problem without performing formal calculations.

Solution

The empty boxes are identical, so they do not contribute to any mass or density difference. Since the edge of the cube and the diameter of the sphere are identical, they will occupy the same volume in each of the boxes; therefore, each box will contain the same number of cubes or spheres. If you view the spheres as cubes that have been rounded by removing wood, you can conclude that the box containing the cubes must have a greater mass of wood; hence, it must have a greater density.

Conceptual Problem 1.26

Consider the following compounds and their densities.

Substance	Density (g/mL)	Substance	Density (g/mL)
isopropyl alcohol	0.785	toluene	0.866
n-butyl alcohol	0.810	ethylene glycol	1.114

You create a column of the liquids in a glass cylinder with the most dense material on the bottom layer and the least dense on the top. You do not allow the liquids to mix.

a. First you drop a plastic bead that has a density of 0.24 g/cm^3 into the column. What do you expect to observe?

b. Next you drop a different plastic bead that has a volume of 0.043 mL and a mass of 3.92 x 10^{-22} g into the column. What would you expect to observe in this case?

c. You drop another bead into the column and observe that it makes it all the way to the bottom of the column. What can you conclude about the density of this bead?

Concept Target

• Use the numerical representation of density to construct a model and employ this model to interpret and predict experimental results.

Solution

a. Since the bead is less dense than any of the liquids in the container, the bead will float on top of all the liquids.

b. First, determine the density of the plastic bead. Since density is mass divided by volume, you get

$$d = \frac{m}{V} = \frac{3.92 \text{ x } 10^{-2} \text{ g}}{0.043 \text{ mL}} = 0.9\underline{1}1 \text{ g/mL} = 0.92 \text{ g/mL}$$

Thus, the glass bead will pass through the top three layers and float on the ethylene glycol layer, which is more dense.

c. Since the bead sinks all the way to the bottom, it must be more dense than 1.114 g/mL.

Conceptual Problem 1.27

a. Which of the following items have a mass of about 1 g?

a grain of sand	a 5.0-gallon bucket of water
a paper clip	a brick
a nickel	a car

b. What is the approximate mass (using SI mass units) of each of the items above?

Concept Target

• Use common objects to improve understanding of the metric quantities used to express mass.

Solution

a. A paper clip has a mass of about 1 g.
b. Answers will vary depending upon your particular sample. Keeping in mind that the SI unit for mass is kg, the approximate weights for the items presented in the problem are: a grain of sand, 1×10^{-5} kg; a paper clip, 1×10^{-3} kg; a nickel, 5×10^{-3} kg; a 5.0-gallon bucket of water, 2.0×10^{1} kg; a brick, 3 kg; a car, 1×10^{3} kg.

Conceptual Problem 1.28

What is the length of the nail reported to the correct number of significant figures?

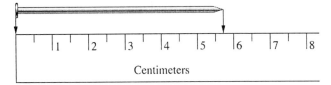

Centimeters

Concept Target

• Obtain the correct number of significant figures when taking measurements.

Solution

When taking measurements, never throw away meaningful information even if there is some uncertainty in the final digit. In this case, you are certain that the nail is between 5 and 6 cm. The uncertain, yet still important digit, is between the 5 and 6 cm measurements. You can estimate, with reasonable precision, that it is about 0.7 cm from the 5 cm mark, so an acceptable answer would be 5.7 cm. Another person might argue that the length of the nail is closer to 5.8 cm, which is also acceptable given the precision of the ruler. In any case, an answer of 5.7 or 5.8 should provide useful information about the length of the nail. If you were to report the length of the nail as 6 cm, this is discarding potentially useful length information provided by the measuring instrument. If a higher degree of measurement precision were needed (more significant figures), you would need to switch to a more precise ruler: for example, one that had mm markings.

Chapter 2

Atoms, Molecules, and Ions

Concept Check 2.1

Like Dalton, chemists continue to model atoms using spheres. Modern models are usually drawn with a computer and use different colors to represent atoms of different elements. Which of the models below represents CO_2?

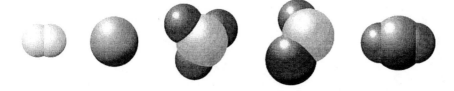

Concept Target

• Apply Dalton's atomic theory to molecular models.

Solution

CO_2 is a compound that is a combination of one carbon atom combined with two oxygen atoms. Therefore, the chemical model must contain a chemical combination of three atoms stuck together, with two of the atoms being the same (oxygen). Since each "ball" represents an individual atom, the three models on the left can be eliminated since they don't contain the correct number of atoms. Keeping in mind that balls of the same color represent the same element, only the model on the far right contains two elements with the correct ratio of atoms, 1:2, therefore it must be CO_2.

Concept Check 2.2

What would be a feasible model for the atom if Geiger and Marsden had found that 7999 out of 8000 alpha particles were deflected back at the alpha-particle source?

Concept Target

• Explore the relationship between experimental results and a model.

Solution

If 7999 out of 8000 alpha particles deflected back at the alpha-particle source, this would imply that an atom is a solid, impenetrable mass. Keep in mind that this is in direct contrast to what was observed in the actual experiments, where the majority of the alpha particles passed through without being deflected.

Concept Check 2.3

Consider the elements He, Ne, and Ar. Can you think of a reason why they are in the same group in the periodic table?

Concept Target

• Explore some of the relationships among elements in the periodic table.

Solution

Elements are listed together in groups because they have similar chemical and/or physical properties.

Concept Check 2.4

Identify the following compounds as being a hydrocarbon, alcohol, ether, or carboxylic acid.

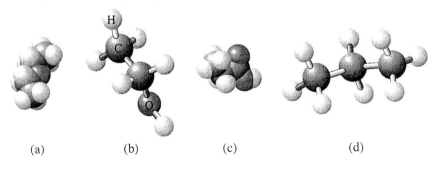

(a)   (b)   (c)   (d)

Concept Target

- Recognize organic functional groups and organic compounds.
- Interpret molecular models.

Solution

a. This compound is an ether because the compound has a functional group of an oxygen atom between two carbon atoms (-O-).
b. This compound is an alcohol because it has an –OH functional group.
c. This compound is a carboxylic acid because it has the –COOH functional group.
d. This compound is a hydrocarbon because it contains only carbon and hydrogen atoms.

Concept Check 2.5

You take a job with the U.S. Environmental Protection Agency (EPA) inspecting college chemistry laboratories. On the first day of the job while inspecting Generic University, you encounter a bottle with the formula Al_2Q_3 that was used as an unknown compound in an experiment. Before you send the compound off to the EPA lab for analysis, you want to narrow down the possibilities for element Q. What are the likely (real element) candidates for element Q?

Concept Target

- Recognize that Al_2Q_3 is likely an ionic compound.
- Determine ionic charges from the periodic table and the chemical formula of a compound.

Solution

A bottle containing a compound with the formula Al_2Q_3 would have an anion Q with a charge of 2-. The total positive charge in the compound due to the Al^{3+} is 6+ (2 x 3+), so the total negative charge must be 6-; therefore, each Q ion must have a charge of 2-. Thus, Q would probably be an element from Group VIA on the periodic table.

Conceptual Problem 2.21

One of the early models of the atom proposed that atoms were wispy balls of positive charge with the electrons evenly distributed throughout. What would you expect to observe if you conducted Rutherford's experiment and the atom had this structure?

Concept Target

• Demonstrate an understanding of the current model of the atom and how the interpretation of experimental data led to this model.

Solution

If atoms were balls of positive charge with the electrons evenly distributed throughout, there would be no massive, positive nucleus to deflect the beam of alpha particles when it is shot at the gold foil.

Conceptual Problem 2.22

A friend is trying to balance the following equation.

$$N_2 + H_2 \rightarrow NH_3$$

He presents you with his version of the "balanced" equation.

$$N + H_2 \rightarrow NH_3$$

You immediately recognize that he has committed a serious error; however, he argues that there is nothing wrong since the equation is balanced. What reason can you give to convince him that his "method" of balancing the equation is flawed?

Concept Target

• Avoid the incorrect technique of changing the subscripts of chemical formulas as a method of balancing chemical equations.

Solution

Once the subscripts of the compounds in the original chemical equation are changed (the molecule N_2 was changed to the atom N), the substances reacting are no longer the same. Your friend may be able to balance the second equation, but it is no longer the same chemical reaction.

Conceptual Problem 2.23

Given that the periodic table is an organizational scheme for the elements, what might be some other logical ways of grouping the elements that would provide meaningful chemical information?

Concept Target

- Emphasize that the periodic table is a convenient organizational scheme, but it is not the only way to organize the elements.

Solution

You could group elements by similar physical properties, such as density, mass, color, conductivity, etc., or by chemical properties, such as reaction with air, reaction with water, etc.

Conceptual Problem 2.24

You discover a new set of polyatomic anions that have the newly discovered element "X" combined with oxygen. Since you made the discovery, you get to name these new polyatomic ions. In developing the names, you want to follow convention. How would you name the following, given that a colleague in the lab came up with a name that she really likes for one of the ions?

Formula	Name
XO_4^{2-}	
XO_3^{2-}	
XO_2^{2-}	excite
XO^{2-}	

Concept Target

- Recognize that once the root name of a polyatomic ion in a group of ions is determined, there is a systematic approach that can be applied to their naming.

Solution

You would name the ions with the formula XO_4^{2-}, XO_3^{2-}, and XO^{2-} using the name for XO_2^{2-} (excite) as the example to determine the root name of the element X (exc). Thus, XO_4^{2-}, with the greatest number of oxygen atoms in the group, would be perexcate; XO_3^{2-} would be excate; and XO^{2-}, with the fewest oxygen atoms in the group, would be hypoexcite.

Conceptual Problem 2.25

You have the mythical metal element "X" that can exist as X^+, X^{2+}, and X^{5+} ions.

a. What would be the chemical formulas for compounds that form from the combination of each of the X ions and SO_4^{2-}?

b.   If the name of the element X is exy, what would be the names of each of the compounds from part a. of this problem?

Concept Target

• Grasp how cations and anions can be combined to form ionic compounds.
• Show that when writing the name of an ionic compound containing a metal capable of multiple charges, the charge must be indicated in the chemical name.

Solution

a.   In each case, the total positive charge and the total negative charge in the compounds must cancel. Therefore, the compounds with the cations X^+, X^{2+}, and X^{5+}, combined with the SO_4^{2-} anion, are X_2SO_4, XSO_4, and $X_2(SO_4)_5$, respectively.
b.   You recognize the fact that whenever a cation can have multiple oxidation states, 1+, 2+, and 5+ in this case, the name of the compound must indicate this charge. Therefore, the names of the compounds in part a. would be exy(I) sulfate, exy(II) sulfate, and exy(V) sulfate, respectively.

Conceptual Problem 2.26

Match the molecular model with the correct chemical formula: CH_3OH, NH_3, KCl, H_2O. (In order to arrive at the correct answer(s), viewing the color text version of the figure associated with this problem is advisable.)

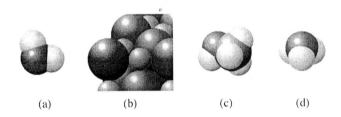

| (a) | (b) | (c) | (d) |

Concept Target

• Connect the chemical formula of a compound to the molecular level representation of the compound.
• Recognize that the formula of an ionic compound represents the ratio of the ions in a solid crystal of that compound.

Solution

The first step in solving this problem is to classify the above compounds as either ionic or molecular (covalent). The only ionic compound is KCl with the others being molecular. Ionic compounds are made up of a crystal composed of a regular three-dimensional arrangement of atoms. Keeping this in mind, moving from models a-d:

a.   This model contains three atoms of two different elements (H and O). Therefore, the model is of H_2O.

b.   This model represents a crystal that contains two different elements in a 1:1 ratio (K^+ and Cl^-). Therefore, the model represents the ionic compound, KCl.

c.   This model contains six atoms, four of which are the same (H), and two others (C and O). Therefore, the model is of CH_3OH.

d.   This model contains four atoms of two different elements (N and H). Therefore, the model is of NH_3.

Conceptual Problem 2.27

Consider a hypothetical case in which the charge on a proton is twice that of an electron. Using this hypothetical case and the fact that atoms maintain a charge of 0, how many protons, neutrons, and electrons would a potassium-39 atom contain?

Concept Target

• Illustrate that the positive and negative charges of protons and electrons "cancel" each other out when combined to form an atom.

Solution

A potassium-39 atom in this case would contain 19 protons and 20 neutrons. If the charge of the proton were twice that of an electron, it would take twice as many electrons as protons, or 38 electrons, to maintain a charge of zero.

Conceptual Problem 2.28

Currently, the atomic mass unit (amu) is exactly 1/12th the mass of a carbon-12 atom and is equal to 1.66×10^{-27} kg.

a.   If the amu were based on sodium-23 with a mass equal to exactly 1/23rd of the mass of a sodium-23 atom, would the mass of the amu be different?

b.   If the new mass of the amu based on sodium-23 is 1.67×10^{-27} kg, how would the mass of a hydrogen atom, in amu, compare with the current mass of a hydrogen atom in amu?

Concept Target

- Show the mass of an atom is not due to the sum of the masses of the protons, neutrons, and electrons: atomic masses are all relative to each other and are based on a defined mass unit called the atomic mass unit (amu).
- Illustrate how changing the defined unit of mass causes the individual masses of each element to change (in amu), but causes no change in the relative mass.

Solution

a.   Since the mass of an atom is not only due to the sum of the masses of the protons, neutrons, and electrons, when you change the element on which you are basing the amu, the mass of the amu must change as well.

b.   Since the amount of material that makes up a hydrogen atom doesn't change, when the amu gets larger, as in this problem, the hydrogen atom must have a smaller mass in amu.

Chapter 3

Calculations with Chemical Formulas and Equations

Concept Check 3.1

You have 1.5 moles of tricycles.
a. How many moles of seats do you have?
b. How many moles of tires do you have?
c. How could you use parts a. and b. as an analogy to teach a friend about the number of moles of OH⁻ ions in 1.5 moles of $Mg(OH)_2$?

Concept Target

• Demonstrate that it is possible to have "moles within moles"; for example, two moles of fish will have four moles of eyes.

Solution

a. Each tricycle has one seat, so you have a total of 1.5 mol of seats.
b. Each tricycle has three tires, so you have 1.5 mol x 3 = 4.5 mol of tires.
c. Each $Mg(OH)_2$ has two OH⁻ ions, so there are 1.5 mol x 2 = 3.0 mol OH⁻ ions.

Concept Check 3.2

You perform combustion analysis on a compound that contains only C and H.
a. Considering the fact that the combustion products CO_2 and H_2O are colorless, how can you tell if some of the product got trapped in the CuO pellets (see Figure 3.6)?

b. Would your calculated results of mass percentage of C and H be affected if some of the combustion products got trapped in the CuO pellets? If your answer is yes, how might your results differ from the expected value for the compound?

Concept Target

• Understand the combustion analysis experiment.
• Perform a nonnumerical interpretation of experimental results from a combustion analysis experiment.

Solution

a. When conducting this type of experiment, you are assuming that all of the carbon and hydrogen show up in the CO_2 and H_2O, respectively. In this experiment, where all of the carbon and hydrogen do not show up, when you analyze the CO_2 for carbon and H_2O for hydrogen, you will find that the weights in the products are less than those in the carbon and hydrogen you started with.

b. Since you collected less carbon and hydrogen than were present in the original sample, the calculated mass percentage will be less than the expected (real) value. For example, say you have a 10.0-g sample that contains 7.5 g of carbon. You run the experiment on the 10.0-g sample and collect only 5.0 g of carbon. The calculated percent carbon based on your experimental results would be 50% instead of the correct amount of 75%.

Concept Check 3.3

A friend has some questions about empirical formulas and molecular formulas. You can assume that he is good at performing the calculations.

a. For a problem that asked him to determine the empirical formula, he found the answer $C_2H_8O_2$. Is this a possible answer to the problem? If not, what guidance would you offer your friend?

b. For another problem he found the answer $C_{1.5}H_4$ as the empirical formula. Is this answer correct? Once again, if it isn't correct, what could you do to help your friend?

c. Since you have been a big help, your friend asks one more question. He completed a problem of the same type as Example 3.11. His answers indicate that the compound has an empirical formula of C_3H_8O and the molecular formula C_3H_8O. Is this result possible?

Concept Target

• Develop a complete understanding of the term "empirical formula."

Solution

a. $C_2H_8O_2$ is not an empirical formula, since each of the subscripts can be divided by two to obtain a possible empirical formula of CH_4O. (The empirical formula is not the <u>smallest</u> integer ratio of subscripts.)

b. $C_{1.5}H_4$ is not a correct empirical formula, because one of the subscripts is not an integer. Multiply each of the subscripts by two to obtain the possible empirical formula C_3H_8. (Since the subscript of carbon is the decimal number 1.5, the empirical formula is not the smallest <u>integer</u> ratio of subscripts.)

c.   Yes, the empirical formula and the molecular formula can be the same, as is the case in this problem where the formula is written with the smallest integer subscripts.

Concept Check 3.4

The main reaction of a charcoal grill is
$$C(s) + O_2(g) \rightarrow CO_2(g)$$
Which of the statements below are incorrect? Why?

a.   1 atom of carbon reacts with 1 molecule of oxygen to produce 1 molecule of CO_2.
b.   1 g of C reacts with 1 g of O_2 to produce 2 g of CO_2.
c.   1 g of C reacts with 0.5 g of O_2 to produce 1 g of CO_2.
d.   12 g of C reacts with 32 g of O_2 to produce 44 g of CO_2.
e.   1 mol of C reacts with 1 mol of O_2 to produce 1 mol of CO_2.
f.   1 mol of C reacts with 0.5 mol of O_2 to produce 1 mol of CO_2.

g.

h.

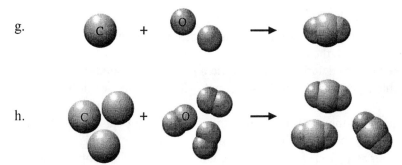

Concept Target

• Correctly interpret balanced chemical equations.

Solution

a. Correct. The coefficients in balanced equations can represent amounts in atoms and molecules.
b. Incorrect. The coefficients in a balanced chemical equation do not represent amounts in grams. One gram of carbon and one gram of oxygen represent different molar amounts.
c. Incorrect. The coefficients in a balanced chemical equation do not represent amounts in grams.
d. Correct. You might initially think that this is an incorrect representation; however, 12 g of C, 32 g of O_2, and 44 g of CO_2 each represent one mole of the substance, so the relationship of the chemical reaction is obeyed.
e. Correct. The coefficients in balanced equations can represent amounts in moles.
f. Incorrect. The amount of O_2 present is not enough to react completely with one mole of carbon. Only one-half of the carbon would react, and one-half mole of CO_2 would form.
g. Incorrect. In this representation, oxygen is being shown as individual atoms of O, not as molecules of O_2, so the drawing is not correctly depicting the chemical reaction.
h. Correct. The molecular models correctly depict a balanced chemical reaction since the same number of atoms of each element appears on both sides of the equation.

Concept Check 3.5

You perform the hypothetical reaction of an element, $X_2(g)$, with another element, $Y(g)$, to produce $XY(g)$.
a. Write the balanced chemical equation for the reaction.
b. If X_2 and Y were mixed in the quantities shown in the container below and allowed to react, which of the following options is the correct representation of the contents of the container after the reaction has occurred?
c. Using the information presented in part b., identify the limiting reactant.

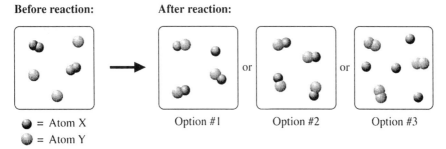

Before reaction: After reaction:

Option #1   Option #2   Option #3

○ = Atom X
○ = Atom Y

Concept Target

• Develop an understanding of the concept of limiting reactants on a molecular level.

Solution

a. $X_2(g) + 2Y(g) \rightarrow 2XY(g)$
b. Since the product consists of a combination of X and Y in a 1:1 ratio, it must consist of two atoms hooked together. If you count the total number of X atoms (split apart the X_2 molecules) and Y atoms present prior to the reaction there are four X atoms and three Y atoms. From these starting quantities, you are limited to three XY molecules and left with an unreacted Y. Option #1 represents this situation and is therefore the correct answer.
c. Since $X_2(g)$ was completely used up during the course of the reaction, it is the limiting reactant.

Conceptual Problem 3.13

You react nitrogen and hydrogen in a container to produce ammonia, $NH_3(g)$. The following figure depicts the contents of the container after the reaction is complete.

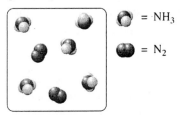

= NH_3

= N_2

a. Write a balanced chemical equation for the reaction.
b. What is the limiting reactant?
c. How many molecules of the limiting reactant would you need to add to the container in order to have a complete reaction (convert all reactants to products)?

Concept Target

• Understand the important concepts associated with chemical reactions on a molecular level.

Solution

a. $3H_2(g) + N_2(g) \rightarrow 2NH_3(g)$
b. Since there is no H_2 present in the container, it was entirely consumed during the reaction which makes it the limiting reactant.
c. According to the chemical reaction, three molecules of H_2 are required for every one molecule of N_2. Since there are two moles of unreacted N_2, you would need six additional moles of H_2 to complete the reaction.

Conceptual Problem 3.14

Propane, C_2H_6, is the fuel of choice in a gas barbecue. When propane burns, the balanced equation is $2C_2H_6 + 7O_2 \rightarrow 4CO_2 + 6H_2O$.
a. What is the limiting reactant when cooking with a gas grill?
b. If the grill will not light and you know that you have an ample flow of propane to the burner, what is the limiting reactant?
c. When using a gas grill you can sometimes turn the gas up to the point at which the flame becomes yellow and smoky. In terms of the chemical reaction, what is happening?

Concept Target

• Explain the results of a common reaction (experiment) in terms of limiting and excess reactants.

Solution

a. The limiting reactant when cooking with a gas grill would be the propane. This makes sense since propane is the material that you must purchase in order to cook your food.
b. Since the chemical reaction only requires propane and oxygen, if the grill will not light with ample propane present, then the limiting reactant must be oxygen.
c. Once again, here is a case where you have adequate propane, so you can conclude that a yellow flame indicates that not enough oxygen is present to combust all of the propane. If there is not enough O_2 available for complete combustion, a reasonable assumption is that some of the products will have fewer oxygen atoms than CO_2. Therefore, a mixture of products would be obtained in this case, including carbon monoxide (CO) and soot (carbon particles).

Conceptual Problem 3.15

A critical point to master in becoming proficient at solving problems is evaluating whether or not your answer is reasonable. A friend asks you to look over her homework to see if she has done the calculations correctly. Shown below are descriptions of some of her answers. Without using your calculator or doing calculations on paper, see if you can judge the answers below as being reasonable or ones that will require her to go back and work the problems again.
a. 0.33 mol of an element has a mass of 1.0×10^{-3} g.
b. The mass of one molecule of water is 1.80×10^{-10} g.
c. There are 3.01×10^{23} atoms of Na in 0.500 mol of Na.
d. The molar mass of CO_2 is 44.0 kg/mol.

Concept Target

- Evaluate numerical answers to common problem types to see if they fit within reasonable limits.

Solution

a. This answer is unreasonable because 1.0×10^{-3} g is too small for 0.33 mole of an element. For example, 0.33 mole of hydrogen, the lightest element, would have a mass of 3.3×10^{-1} g.

b. This answer is unreasonable because 1.80×10^{-10} g is too large for one water molecule. (The mass of one water molecule is 2.99×10^{-23} g.)

c. This answer is reasonable because 3.01×10^{23} is one-half of Avogadro's number.

d. This answer is unreasonable because the units for molar mass should be g/mol, so this quantity is 1000 times too large.

Conceptual Problem 3.16

An exciting, and often loud, chemical demonstration involves the simple reaction of hydrogen gas and oxygen gas to produce water vapor: $2H_2(g) + O_2(g) \rightarrow 2H_2O(g)$. The reaction is carried out in soap bubbles or balloons filled with the reactant gases. You get the reaction to proceed by igniting the bubbles or balloons. The more H_2O that is formed during the reaction, the bigger the bang. Explain the following observations.

a. A bubble containing just H_2 makes a quiet "fffft" sound when ignited.

b. When a bubble containing equal amounts of H_2 and O_2 is ignited, a sizable bang results.

c. When a bubble containing a ratio of 2 to 1 in the amounts of H_2 and O_2 is ignited, the loudest bang results.

d. When a bubble containing just O_2 is ignited, virtually no sound is made.

Concept Target

- Relate experimental results to the stoichiometry of a balanced chemical equation (limiting reactants).

Solution

a. In order to have a complete reaction, a ratio of two moles of hydrogen to every mole of oxygen is required. In this case, there is not enough oxygen in the air outside of the bubble for the complete reaction of hydrogen.

b. In this case you have a ratio of one mole of H_2 to one mole of O_2. According to the balanced chemical reaction, every one mole of O_2 can react with two moles of H_2. In this case, when 0.5 mole of O_2 has reacted, all of the H_2 (one mole) will be consumed, leaving behind 0.5 mole of unreacted O_2.

c. In this case you have a ratio of two moles of H_2 to one mole of O_2, which is the correct stoichiometric amount, so all of the hydrogen and all of the oxygen react completely.

d. In order for the reaction to occur, <u>both</u> oxygen and hydrogen must be present. Oxygen does not combust, and there is no hydrogen present to burn so no reaction occurs.

Conceptual Problem 3.17

High cost and limited availability of a reactant often dictate which reactant is limiting in a particular process. Identify the limiting reactant when running the reactions below and give a reason to support your decision.

a. Burning charcoal on a grill: $C(s) + O_2(g) \rightarrow CO_2(g)$
b. Burning a chunk of Mg in water: $Mg(s) + 2H_2O(l) \rightarrow Mg(OH)_2(aq) + H_2(g)$
c. The Haber process of ammonia production: $3H_2(g) + N_2(g) \rightarrow 2NH_3(g)$

Concept Target

• Use common sense, everyday observations, and information from the text to determine limiting reactants in chemical equations.

Solution

a. The limiting reactant would be the charcoal because the air would supply as much oxygen as needed.

b. The limiting reactant would be the magnesium because the beaker would contain much more water than is needed for the reaction (approximately 18 mL of water is one mole).

c. The limiting reactant would be the H_2 because the air could supply as much nitrogen as is needed.

Conceptual Problem 3.18

A few hydrogen and oxygen molecules are introduced into a container in the quantities depicted in the following drawing. The gases are then ignited by a spark causing them to react and form H_2O.

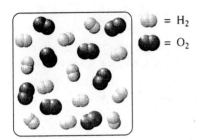

= H$_2$

= O$_2$

a. What is the maximum number of water molecules that can be formed in the chemical reaction?

b. Draw a molecular level representation of the container's contents after the chemical reaction.

Concept Target

• Employ molecular models to present the important concepts associated with chemical reactions.

Solution

a. Since the balanced chemical equation for the reaction is $2H_2 + O_2 \rightarrow 2H_2O$, in order to form the water, you need two molecules of hydrogen for every one molecule of oxygen. Given the quantities of reactants present in the container (12 H$_2$ molecules and 8 O$_2$ molecules) and applying the 2:1 ratio, you can produce a maximum of 12 molecules of water.

b. The drawing of the container after the reaction should contain 12 H$_2$O molecules and two O$_2$ molecules.

Chapter 4

Chemical Reactions: An Introduction

Concept Check 4.1

LiI(s) and CH$_3$OH(l) are each introduced into a separate beaker containing water. Using the drawings shown here, label each beaker with the appropriate compound and indicate whether you would expect each substance to be a strong electrolyte, weak electrolyte, or nonelectrolyte. (In order to arrive at the correct answer(s), viewing the color text version of the figure associated with this problem is advisable.)

Concept Target

- Develop a molecular view of solutions that are electrolytes and nonelectrolytes.
- Apply the solubility rules for ionic compounds.

Solution

The left beaker contains two types of individual atoms (ions) and no solid, therefore it must represent the soluble, LiI. Because Li is a soluble ionic compound, it is an electrolyte. The beaker on the right represents a molecular compound that is soluble but not dissociated in solution. Therefore, it must be the CH$_3$OH. Because the CH$_3$OH is not dissociated in solution and no ions are present, it is a nonelectrolyte.

Concept Check 4.2

Your lab partner tells you that she mixed two solutions that contain ions. You analyze the solution and find that it contains the ions and precipitate shown in the beaker.

$Na^+(aq)$

$C_2H_3O_2^-(aq)$

$SrSO_4(s)$

a. Write the molecular equation for the reaction.
b. Write the complete ionic equation for the reaction.
c. Write the net ionic equation for the reaction.

Concept Target

• Use the data of a completed precipitation reaction presented in a picture to write a molecular equation, a complete ionic equation, and a net ionic equation.

Solution

a. In order to solve this part of the problem, keep in mind that this is an exchange (metathesis) reaction. Since you are given the products in the picture, you need to work backward to determine the reactants. Starting with the solid $SrSO_4(s)$, you know that the SO_4^{2-} anion started the reaction with a different cation (not Sr^{2+}). Since Na^+ is the only option, you can conclude that one of the reactants must be Na_2SO_4. Based on solubility rules, you know that the Na_2SO_4 is soluble, so you represent it as $Na_2SO_4(aq)$. The remaining cation and anion indicate that the other reactant is the soluble $Sr(C_2H_3O_2)_2$. Observing the soluble and insoluble species in the picture, you conclude that the molecular equation is

$$Na_2SO_4(aq) + Sr(C_2H_3O_2)_2(aq) \rightarrow SrSO_4(s) + 2NaC_2H_3O_2(aq)$$

b. Writing the strong electrolytes in the form of ions and the solid with its molecular formula, the complete ionic equation for the reaction is

$$2Na^+(aq) + SO_4^{2-}(aq) + Sr^{2+}(aq) + 2C_2H_3O_2^-(aq) \rightarrow SrSO_4(s) + 2Na^+(aq) + 2C_2H_3O_2(aq)$$

c. After canceling the spectator ions, the net ionic equation for the reaction is

$$Sr^{2+}(aq) + SO_4^{2-}(aq) \rightarrow SrSO_4(s)$$

Concept Check 4.3

At times, we want to generalize the formulas of certain important chemical substances: acids and bases fall into this category. Given the following reactions, try to identify the acids, bases, and some examples of what the general symbols (M and A⁻) represent.

a. $MOH(s) \rightarrow M^+(aq) + OH^-(aq)$
b. $HA(aq) + H_2O(l) \leftrightarrows H_3O^+(aq) + A^-(aq)$
c. $H_2A(aq) + H_2O(l) \leftrightarrows H_3O^+(aq) + HA^-(aq)$
d. For parts a. to c., give real examples for M and A.

Concept Target

• Recognize the formulas and reactions of acids and bases (weak and strong).
• Provide an introduction to the generic representation of some bases, monoprotic acids, and diprotic acids.

Solution

a. MOH must be a base since OH⁻ is being produced in solution. It is a strong base because the reaction indicates that MOH is completely soluble (strong electrolyte). In order to maintain charge balance in the formula, the element M must be a 1+ cation, probably a metal from Group IA of the periodic table. Examples of bases that fall into this category include NaOH and KOH.
b. This must be an acid since H⁺ is being produced in solution. It is a weak acid because the double arrow is used, indicating only partial ionization in solution. From the chemical reaction, A⁻ represents an anion with a 1- charge. Acetic acid, $HC_2H_3O_2$, is a weak acid of this type.
c. This must be an acid since H⁺ is being produced in solution. $H_2A(aq)$ is a weak acid because the equation indicates only partial ionization in solution. A represents an anion with a 2- charge. Carbonic acid, H_2CO_3, is a weak acid of this type.
d. Examples of M include Na^+, K^+, and Li^+. Examples of A for reaction b. include F^-, $C_2H_3O_2^-$, and CN^-. Examples of A for reaction c. include S^{2-}, CO_3^{2-}, and $C_4O_4O_6^{2-}$.

Concept Check 4.4

Consider the following beakers. Each contains a solution of the hypothetical atom X.

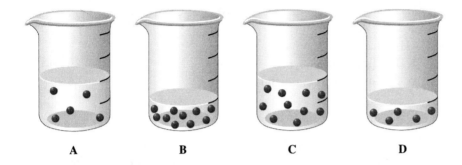

a. Arrange the beakers in order of increasing concentration of X.
b. Without adding or removing X, what specific things could you do to make the concentrations of X equal in each beaker? (*Hint:* Think about dilutions.)

Concept Target

• Develop a molecular level understanding of concentration and dilution.

Solution

a. In order to answer this question, you need to compare the number of atoms of X per unit of volume. In order to compare volumes, use the lines on the sides of the beakers. Beaker A has a concentration of 5 atoms per 2 volume units, 5/2 or 2.5/1. Beaker B has a concentration of 10 atoms per 1 volume unit, 10/1. Beaker C has a concentration of 10 atoms per 2 volume units, 10/2 or 5/1. Beaker D has a concentration of five atoms per volume unit, 5/1. Comparing the concentrations, the ranking from highest to lowest concentration is: Beaker B > Beaker C = Beaker D > Beaker A.
b. To make the concentrations of X equal in each beaker, they all have to be made to match the beaker with the lowest concentration. This is Beaker A, which has 5 atoms of X in one-half a beaker of solution. To make the concentrations equal, do the following: double the volume of Beakers C and D, and quadruple the volume of Beaker B. Overall, Beakers A and B will contain a full beaker of solution, and Beakers C and D will contain a half-beaker of solution.

Concept Check 4.5

Consider three flasks, each containing 0.10 mol of acid. You need to learn something about the acids in each of the flasks, so you perform titration using an NaOH solution. Here are the results of the experiment:

Flask A   10 mL of NaOH required for neutralization
Flask B   20 mL of NaOH required for neutralization
Flask C   30 mL of NaOH required for neutralization

a. What have you learned about each of these acids from performing the experiment?
b. Could you use the results of this experiment to determine the concentration of the NaOH? If not, what assumption about the molecular formulas of the acids would allow you to make the concentration determination?

Concept Target

• Relate the results of a titration experiment to the chemical structure of acids dissolved in solution.
• Understand the stoichiometry of acid-base reactions.

Solution

a. Since flask C required three times the amount of titrant (NaOH) as acid A, you have learned that acid C has three times as many acidic protons as acid A. Since flask B required two times the amount of titrant as acid A, you have also learned that acid B has two times as many acidic protons as acid A.
b. If you assume that acid A contains a monoprotic acid, then you know the number of moles of A in the flask. After performing the titration, you know that the moles of NaOH must equal the moles of acid in flask A. You take the number of moles of NaOH and divide it by the volume of NaOH added during the titration to determine the concentration of the NaOH solution.

Conceptual Problem 4.15

You need to perform gravimetric analysis of a water sample in order to determine the amount of Ag^+ present.
a. List three aqueous solutions that would be suitable for mixing with the sample to perform the analysis.
b. Would adding $KNO_3(aq)$ allow you to perform the analysis?
c. Assume you have performed the analysis and the silver solid that formed is moderately soluble. How might this affect your results?

Concept Target

• Use solubility rules to interpret experimental results.

Solution

a. Any soluble salt that will form a precipitate when reacted with Ag^+ ions in solution will work, for example: $CaCl_2$, Na_2S, $(NH_4)_2CO_3$.
b. No, no precipitate would form.
c. You would underestimate the amount of silver present in the solution.

Conceptual Problem 4.16

In this problem you need to draw two pictures of solutions in beakers at different points in time. Time zero ($t = 0$) will be the hypothetical instant at which the reactants dissolve in the solution (*if* they dissolve) *before* they react. Time after mixing ($t > 0$) will be the time required to allow sufficient interaction of the materials. For now, we assume that insoluble solids have no ions in solution and do not worry about representing the stoichiometric amounts of the dissolved ions. Here is an example.

Solid NaCl and solid $AgNO_3$ are added to a beaker containing 250 mL of water. Note that we are not showing the H_2O and we are representing only the ions and solids in solution.

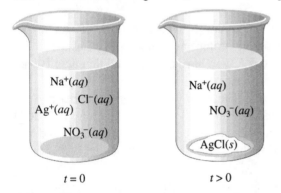

$t = 0$ $t > 0$

Using the same conditions as the example (adding the solids to H_2O), draw pictures of the following:
a. solid lead(II) nitrate and solid ammonium chloride at $t = 0$ and $t > 0$
b. FeS(s) and $NaNO_3$(s) at $t = 0$ and $t > 0$
c. solid lithium iodide and solid sodium carbonate at $t = 0$ and $t > 0$

Concept Target

• Use solubility rules to predict the products of chemical reactions.

- Present the results of solution experiments in a visual form (creating "pictures" of the reactions from written descriptions).

Solution

a.

Left beaker ($t = 0$): $NH_4^+(aq)$, $Cl^-(aq)$, $Pb^{2+}(aq)$, $NO_3^-(aq)$

Right beaker ($t > 0$): $NH_4^+(aq)$, $NO_3^-(aq)$, $PbCl_2(s)$

b.

Left beaker ($t = 0$): $Na^+(aq)$, $NO_3^-(aq)$, $FeS(s)$

Right beaker ($t > 0$): $Na^+(aq)$, $NO_3^-(aq)$, $FeS(s)$

c.

Left beaker ($t = 0$): $Li^+(aq)$, $Na^+(aq)$, $I^-(aq)$, $CO_3^{2-}(aq)$

Right beaker ($t > 0$): $Li^+(aq)$, $Na^+(aq)$, $I^-(aq)$, $CO_3^{2-}(aq)$

Conceptual Problem 4.17

You come across a beaker that contains water, aqueous ammonium acetate, and a precipitate of calcium phosphate.
a. Write the balanced molecular equation for a reaction between two solutions containing ions that could produce this solution.
b. Write the complete ionic equation for the reaction in part a.
c. Write the net ionic equation for the reaction in part a.

Concept Target

• Use experimental data from a chemical reaction to write molecular equations, complete ionic equations, and net ionic equations.
• Apply the solubility rules.

Solution

a. $3Ca(C_2H_3O_2)_2(aq) + 2(NH_4)_3PO_4(aq) \rightarrow Ca_3(PO_4)_2(s) + 6NH_4C_2H_3O_2(aq)$
b. $3Ca^{2+}(aq) + 6C_2H_3O_2^-(aq) + 6NH_4^+(aq) + 2PO_4^{3-}(aq) \rightarrow Ca_3(PO_4)_2(s) + 6NH_4^+(aq) + 6C_2H_3O_2^-(aq)$
c. $3Ca^{2+}(aq) + 2PO_4^{3-}(aq) \rightarrow Ca_3(PO_4)_2(s)$

Conceptual Problem 4.18

Three acid samples are prepared for titration by 0.01 M NaOH:
Sample 1 is prepared by dissolving 0.01 mol of HCl in 50 mL of water.
Sample 2 is prepared by dissolving 0.01 mol of HCl in 60 mL of water.
Sample 3 is prepared by dissolving 0.01 mol of HCl in 70 mL of water.
a. Without performing a formal calculation, compare the concentrations of the three acid samples (rank them from highest to lowest).
b. When performing the titration, which sample, if any, will require the largest volume of the 0.01 M NaOH for neutralization?

Concept Target

• Interpret the results of a titration experiment.
• Demonstrate that when a titration is performed, the concentration of the solution being titrated is not relevant; only the number of moles of material present in the solution being titrated matters.

Solution

a. Sample 1, Sample 2, Sample 3
b. They all will require the same volume of 0.1 M NaOH.

Conceptual Problem 4.19

Would you expect a precipitation reaction between an ionic compound that is an electrolyte and an ionic compound that is a nonelectrolyte? Justify your answer.

Concept Target

• Use information from the terms electrolyte and nonelectrolyte to predict the outcome of chemical reactions.

Solution

Probably not, since the ionic compound that is a nonelectrolyte is not soluble.

Conceptual Problem 4.20

Equal quantities of the hypothetical strong acid HX, weak acid HA, and weak base BZ, are each added to a separate beaker of water, producing the solutions depicted in the drawings. In the drawings, the relative amounts of each substance present in the solution (neglecting the water) are shown. Identify the acid or base that was used to produce each of the solutions (HX, HA, or BZ). (In order to arrive at the correct answer(s), viewing the color text version of the figure associated with this problem is advisable.)

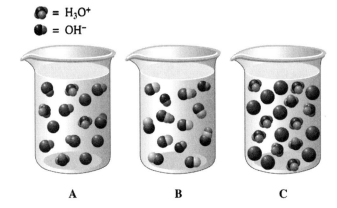

$\bullet$ = H_3O^+
$\bullet$ = OH^-

A   B   C

Concept Target

- Identify the principal difference between a base and an acid.
- Make the distinction between strong and weak acids and bases.

Solution

A good starting point is to identify the solution that contains the base. Since bases produce OH^- in aqueous solution, we would expect to see OH^- present in the BZ solution. The center beaker depicts OH^- in the solution so it must be the base. By default, the remaining two beakers must contain acid. This is confirmed by the presence of H_3O^+ in both the left and right beakers. Keeping in mind that weak acids only partially dissociate, for the weak acid HA, we would expect to observe HA, H_3O^+, and A^- in the solution. In the case of the strong acid, HX, that completely dissociates, we would expect to observe only H_3O^+ and X^- in the solution. The beaker on the right only has H_3O^+ and one other species in solution so it must be the strong acid HX. Examining the beaker on the left, there are three species present, which indicates that it must be the weak acid.

Chapter 5

The Gaseous State

Concept Check 5.1

Suppose that you set up two barometers like the one shown in Figure 5.2. In one of the barometers you use mercury, and in the other you use water. Which of the barometers would have a higher column of liquid, the one with Hg or H_2O? Explain your answer.

Concept Target

• Illustrate that the height of a column of liquid in a barometer is inversely related to the density of the liquid.

Solution

The general relationship between pressure (P) and the height of a liquid column in a barometer is $P = gdh$, where g is the constant acceleration of gravity and d is the density. Examination of this relationship indicates that for a given pressure, as the density of the liquid in the barometer decreases, the height of the liquid must increase. In order to make this relationship more apparent, you can rearrange the equation to:

$$gh = \frac{P}{d}$$

Keeping in mind that you are conducting the experiment at a constant pressure and that gravity is a constant, this mathematical relationship demonstrates that the height of the liquid in the barometer is inversely proportional to the density of the liquid in the barometer.

$$(h \propto \frac{1}{d})$$

This inverse relationship means that as the height of the liquid decreases, the density of the liquid must increase. Since the density of mercury is greater than the density of water, the barometer with the water will have the higher column.

Concept Check 5.2

To conduct some experiments, a 10.0-L flask equipped with a movable plunger, as illustrated here, is filled with enough H_2 gas to cause a pressure of 20 atm.

a. In the first experiment, we decrease the temperature in the flask by 10°C and then increase the volume. Predict how the pressure in the flask changes during each of these events and, if possible, how the final pressure compares with your starting pressure.

b. Once again we start with the pressure in the flask at 20 atm. The flask is then heated 10°C followed by a volume decrease. Predict how the pressure in the flask changes during each of these events and, if possible, how the final pressure compares with your starting pressure.

Concept Target

• Qualitatively explore the relationships between pressure, temperature, and volume of a gas.

Solution

a. In the first step, when the temperature decreases, the pressure will also decrease. This is because, according to the combined gas law, the pressure is directly proportional to the temperature ($P \propto T$). In the second step, when the volume increases, the pressure will decrease, since according to Boyle's law, pressure and volume are inversely related ($P \propto 1/V$). Both changes result in the pressure decreasing, so the final pressure will be less than the starting pressure.

b. In the first step, when the temperature increases, the pressure will also increase. This is because, according to the combined gas law, pressure is directly proportional to the temperature ($P \propto T$). In the second step, when the volume decreases, the pressure will increase, since according to the ideal gas law, pressure and volume are inversely related ($P \propto 1/V$). Both changes result in the pressure increasing, so the final pressure will be greater than the starting pressure.

Concept Check 5.3

Three 3.0-L flasks, each at a pressure of 878 mmHg, contain He, Ar, and Xe.

a. Which of the flasks contains the most atoms of gas?

b. Which of the flasks has the greatest density of gas?

c. If the He flask was heated and the Ar flask was cooled, which of the three flasks would be at the highest pressure?

d. If the temperature of the He was lowered while the Xe was raised, which of the three flasks would have the greatest number of moles of gas?

Concept Target

• Qualitatively apply Avogadro's law and the ideal gas law.

Solution

a. According to Avogadro's law, equal volumes of any two (or more) gases at the same temperature and pressure contain the same number of molecules (or atoms in this case). Therefore, all three flasks contain the same number of atoms.
b. Since density is mass divided by volume, and all three flasks have the same volume (3.0 L), the gas with the largest molar mass, xenon (Xe), will have the greatest density.
c. According to the ideal gas law, $PV = nRT$, pressure is directly proportional to the temperature. Since the helium flask is being heated, it will have the highest pressure.
d. Since the three flasks started with the same number of atoms, and hence the same number of moles, they would all still have the same number of moles no matter how the temperature of the flasks changed.

Concept Check 5.4

A flask equipped with a valve contains 3.0 mol of H_2 gas. You introduce 3.0 mol of Ar gas into the flask via the valve and then seal the flask.
a. What happens to the pressure of just the H_2 gas in the flask after the introduction of the Ar? If it changes, by what factor does it do so?
b. How do the pressures of the Ar and the H_2 in the flask compare?
c. How does the total pressure in the flask relate to the pressures of the two gases?

Concept Target

• Qualitatively interpret Dalton's law of partial pressure and the ideal gas law.

Solution

a. In a mixture of gases, each gas exerts the pressure it would exert if it were the only gas in the flask. The pressure of H_2 is the same whether it is in the flask by itself or with the Ar. Therefore, the pressure of H_2 does not change.
b. According to the ideal gas law, $PV = nRT$, pressure (P) is directly proportional to the number of moles (n). Since the number of moles of H_2 and the number of moles of Ar are equal, their pressures are also equal.

c. The total pressure is equal to the sum of the pressures of the H_2 gas and the Ar gas in the container. The total pressure will also be equal to twice the pressure of the H_2 gas when it was in the flask by itself. It is also equal to twice the pressure that the Ar gas would exert if it were in the flask by itself.

Concept Check 5.5

Consider the following experimental apparatus.

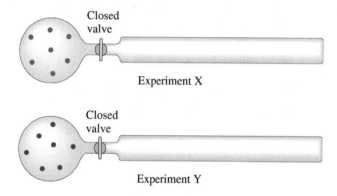

In this setup, each round flask contains a gas and the long tube contains no gas (that is, it is a vacuum).

a. We use 1.0 mol of He for experiment X and 1.0 mol of Ar for experiment Y. If both valves are opened at the same time, which gas would you expect to reach the end of the long tube first?

b. If you wanted the Ar to reach the end of the long tube at the same time as the He, what experimental condition (that is, you cannot change the equipment) could you change to make this happen?

Concept Target

• Use the relationships presented in Graham's law of effusion to interpret experimental results and predict the outcome of experiments.

Solution

a. The rate of effusion is inversely proportional to the square root of the molecular weight of the gas at constant temperature and pressure. Thus, He (molecular weight 4.00 amu) will diffuse faster than Ar (molecular weight 39.95 amu) and reach the end of the tube first.

b. The speed of an atom is directly proportional to the absolute temperature. If you raise the temperature of the Ar, you can make it reach the end of the tube at the same time as the He.

Concept Check 5.6

A 1.00-L container is filled with an ideal gas, and the recorded pressure is 350 atm. We then put the same amount of a real gas into the container and measure the pressure.
a. If the real gas molecules occupy a relatively small volume and have large intermolecular attractions, how would you expect the pressures of the two gases to compare?
b. If the real gas molecules occupy a relatively large volume and there are negligible intermolecular attractions, how would you expect the pressures of the two gases to compare?
c. If the real gas molecules occupy a relatively large volume and have large intermolecular attractions, how would you expect the pressures of the two gases to compare?

Concept Target

• Illustrate the conditions where treating a gas as an ideal gas do not apply.
• Compare the behavior and properties of real and ideal gases under the same conditions.

Solution

a. If the real gas molecules occupy a relatively small volume, then the volume of the gas is essentially equal to the volume of the container, the same as for an ideal gas. However, if there were large intermolecular attractions, the pressure would be less than for an ideal gas. Therefore, the pressure would be greater for the ideal gas.
b. If the real gas molecules occupy a relatively large volume, then the volume available for the gas is less than for an ideal gas, and the pressure would be greater. If there are negligible intermolecular attractions, then the pressure is essentially the same as for an ideal gas. Overall, the pressure would be less for the ideal gas.
c. Since the effects of molecular volume and intermolecular attractions on the pressure of a real gas are opposite, you cannot determine how the pressures of the two gases compare.

Conceptual Problem 5.23

Using the concepts developed in this chapter, explain the following observations.
a. Automobile tires are flatter on cold days.
b. You are not supposed to dispose of aerosol cans in a fire.
c. The lid of a water bottle pops off when the bottle sits in the sun.
d. A balloon pops when you squeeze it.

Concept Target

• Apply the gas laws to everyday observations.

Solution

a. The volume of the tire and the amount of air in the tire remain constant. From the ideal gas law, $PV = nRT$, under these conditions the pressure will vary directly with the temperature ($P \propto T$). Thus, on a cold day, you would expect the pressure in the tires to decrease, and they would appear flatter.

b. Aerosol cans are filled with a fixed amount of gas in a constant volume. From the ideal gas law, under these conditions the pressure will vary directly with the temperature ($P \propto T$). If you put an aerosol can in a fire, you will increase the temperature, and thus the pressure. If the pressure gets high enough, the can will explode.

c. As the water bottle sits in the sun, the liquid water warms up. As the temperature of the water increases, so does its vapor pressure (Table 5.6). If the pressure gets high enough, it will pop the lid off the bottle.

d. The amount of air in the balloon and the temperature remain constant. From the ideal gas law under these conditions, the pressure is inversely proportional to the volume ($P \propto 1/V$). Thus, as you squeeze the balloon, you decrease the volume, resulting in an increase in pressure. If you squeeze hard enough and make the volume small enough, the balloon will pop.

Conceptual Problem 5.24

You have three identical flasks, each containing an equal amount of N_2, O_2, or He. The volume of the N_2 flask is doubled, of the O_2 flask is halved, and of the He flask is reduced to one-third of the original. Rank the flasks from highest to lowest pressure both before and after the volume is changed, and indicate by what factor the pressure has changed.

Concept Target

• Use the combined gas law to qualitatively interpret experimental results.

Solution

Since each of the flasks is identical and each contains an equal amount of gas, the initial pressure in the N_2 flask, the O_2 flask, and the He flask will be the same. After the changes, the pressure in the He flask would be highest, with a pressure equal to three times the original. Next would be the O_2 flask, with a pressure equal to two times the original. Last would be the N_2 flask, with a pressure equal to one-half the original.

Conceptual Problem 5.25

Consider the following glass container equipped with a movable piston.

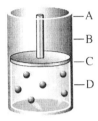

a. By what factor (increase by 1, decrease by 1.5, etc.) would you change the pressure if you wanted the volume to change from volume C to volume D?
b. If the piston were moved from volume C to volume A, by what factor would the pressure change?
c. By what factor would you change the temperature in order to change from volume C to volume B?
d. If you increased the number of moles of gas in the container by a factor of 2, by what factors would the pressure and the volume change?

Concept Target

• Develop a qualitative understanding of the gas law relationships.

Solution

a. The pressure and volume of a gas are inversely proportional; therefore, an increase by a factor of 2 in pressure would decrease the volume by ½ (C to D).
b. The pressure and volume of a gas are inversely proportional; therefore, a decrease by a factor of 2 in pressure would double the volume (C to A).
c. The volume and temperature of a gas are directly proportional; therefore, an increase in kelvin temperature by a factor of 1.5 would result in an increase in volume by a factor of 1.5 (C to B).
d. Since the piston can move, the pressure would not change (it would be equal to the starting pressure). The volume of gas is directly proportional to the number of moles; therefore an increase in the number of moles by a factor of 2 would cause the volume to increase by 2 (C to A).

Conceptual Problem 5.26

A 3.00-L flask containing 2.0 mol of O_2 and 1.0 mol of N_2 is in a room with a temperature of 22.0°C.

a. How much (what fraction) of the total pressure in the flask is due to the N_2?
b. The flask is cooled and the pressure drops. What happens, if anything, to the mole fraction of the O_2 at the lower temperature?
c. 1.0 L of liquid water is introduced into the flask containing both gases. The pressure is measured about 45 minutes later. Would you expect the measured pressure to be higher or lower?
d. Given the information in this problem and the conditions in part c., would it be possible to calculate the pressure in the flask after the introduction of the water? If it is not possible with the given information, what further information would you need to accomplish this task?

Concept Target

• Develop an integrated understanding of partial pressure, mole fraction, and the vapor pressure of liquids.

Solution

a. Since 1.0 out of the 3.0 moles of gas in the container is N_2, the fraction of the pressure due to N_2 is 1/3.
b. Mole fractions are not a function of temperature, so nothing would happen.
c. You would expect the pressure to be higher for two reasons. First, the water would occupy some volume, reducing the volume available for the gas to occupy. Thus, according to Boyle's law, as volume decreases, pressure increases. Second, after a time, the water would evaporate, and the vapor pressure due to the water would contribute to the total pressure, thereby increasing it.
d. Yes, there is enough information in the problem to calculate the pressure in the flask, but you would also need to know the vapor pressure of water at 22.0°C (Table 5.6).

Conceptual Problem 5.27

Consider the following setup, which shows identical containers connected by a tube with a valve that is presently closed. The container on the left has 1.0 mol of H_2 gas; the container on the right has 1.0 mol of O_2.

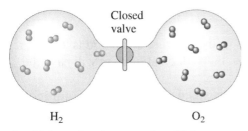

H_2 O_2

Note: Acceptable answers to some of these questions might be "both" or "neither one."
a. Which container has the greatest density of gas?
b. Which container has molecules that are moving at a faster average molecular speed?
c. Which container has more molecules?

d. If the valve is opened, will the pressure in each of the containers change? If it does, how will it change (increase, decrease, or no change)?
e. 2.0 mol of Ar is added to the system with the valve open. What fraction of the total pressure will be due to the H_2?

Concept Target

• Integrate concepts from the ideal gas law, Avogadro's law, Dalton's law of partial pressure, and the kinetic molecular theory.

Solution

a. The container with the O_2 has the greater density, since the molar mass of O_2 (32.00 g/mol) is greater than for H_2 (2.016 g/mol).
b. Since the H_2 molecules are lighter, they will be moving faster.
c. Both containers have the same number of molecules (Avogadro's law).
d. The pressure in each of the containers will not change when the valve is opened. Each container starts with the same pressure. Since the total volume remains constant, the pressure will not change.
e. The fraction of the total pressure due to the H_2 would now be 1/4.

Conceptual Problem 5.28

Two identical He-filled balloons, each with a volume of 20 L, are allowed to rise into the atmosphere. One rises to an altitude of 3000 m while the other rises to 6000 m.
a. Assuming that the balloons are at the same temperature, which balloon has the greater volume?
b. What information would you need to calculate the volume of each balloon at its respective height?

Concept Target

• Qualitatively apply Boyle's law.

Solution

a. Pressure decreases as you increase in altitude. Thus, the pressure at 6000 m is less than the pressure at 3000 m. For two identical balloons, the balloon at 6000 m will have the greater volume (Boyle's law).
b. In order to calculate the volume of each balloon, you would need the temperature and pressure on the ground and the temperature and pressure at their respective heights.

Conceptual Problem 5.29

You have a balloon that contains O_2. What could you do to the balloon in order to double the volume? Be specific in your answers; for example, you could increase the number of moles of O_2 by a factor of two.

Concept Target

• Ensure a good qualitative understanding of the gas laws.

Solution

In order to double the volume you could reduce the pressure by 1/2. You could also increase the temperature, but you cannot determine the final temperature without knowing the initial temperature.

Conceptual Problem 5.30

Three 25.0-L flasks are placed next to each other on a shelf in a chemistry stockroom. The first flask contains He at a pressure of 1.0 atm, the second contains Xe at 1.50 atm, the third contains F_2 and has a label that says 2.0 mol F_2. Consider the following questions about these flasks.
a. Which flask has the greatest number of moles of gas?
b. If you wanted each of the flasks to be at the same pressure as the He flask, what general things could you do to the other two containers to make this happen?

Concept Target

• Ensure a good qualitative understanding of the gas laws.

Solution

a. If you assume that the flasks are at ordinary room temperature, say 25°C, then there would be approximately one mole of He (at 1.0 atm) and 1.5 mol of Xe (at 1.5 atm). Thus, the F_2 flask (with 2.0 mol) would contain the greatest number of moles of gas.
b. You could either decrease the volume, increase the temperature, or both.

Chapter 6

Thermochemistry

Concept Check 6.1

A solar-powered water pump has photovoltaic cells on protruding top panels. These cells collect energy from sunlight, storing it momentarily in a battery, which later runs an electric motor that pumps water up to a storage tank on a hill. What energy conversions are involved in using sunlight to pump water into the storage tank?

Concept Target

• Give concrete understanding to the law of conservation of energy by describing the specific energy changes that occur in a particular instance.

Solution

The photovoltaic cells collect the sun's energy, converting it to electric energy. This electric energy is stored in the battery as chemical energy, which is later changed back to electric energy that runs a motor. As the motor rotates, it changes the electric energy to kinetic energy (energy of motion) of the motor, then of the water, which in turn is changed to potential energy (energy of position) of water as the water moves upward in the gravitation field of earth.

Concept Check 6.2

Natural gas consists primarily of methane, CH_4. It is used in a process called steam reforming to prepare a gaseous mixture of carbon monoxide and hydrogen for industrial use.

$$CH_4(g) + H_2O(g) \rightarrow CO(g) + 3H_2(g); \Delta H = 206 \text{ kJ}$$

The reverse reaction, the reaction of carbon monoxide and hydrogen, has been explored as a way to prepare methane (synthetic natural gas). Which of the following are exothermic? Of these, which one is the most exothermic?

a. $CH_4(g) + H_2O(g) \rightarrow CO(g) + 3H_2(g)$
b. $2CH_4(g) + 2H_2O(g) \rightarrow 2CO(g) + 6H_2(g)$
c. $CO(g) + 3H_2(g) \rightarrow CH_4(g) + H_2O(g)$
d. $2CO(g) + 6H_2(g) \rightarrow 2CH_4(g) + 2H_2O(g)$

Concept Target

• Think about the relationships among heat, enthalpy change, and thermochemical equations.

Solution

a. This reaction is the one shown in the problem, and it has a positive ΔH, so the reaction is endothermic.
b. This reaction is simply twice a., so it is also endothermic.
c. This reaction is the reverse of a., so it is exothermic.
d. This reaction is simply twice that of c., so it is more exothermic than c.
Thus, d. is the most exothermic reaction.

Concept Check 6.3

The heat of fusion (also called heat of melting), ΔH_{fus}, of ice is the enthalpy change for
   $H_2O(s) \rightarrow H_2O(l); \ \Delta H_{fus}.$
Similarly, the heat of vaporization, ΔH_{vap}, of liquid water is the enthalpy change for
   $H_2O(l) \rightarrow H_2O(g); \ \Delta H_{vap}.$
How is the heat of sublimation, ΔH_{sub}, the enthalpy change for the reaction
   $H_2O(s) \rightarrow H_2O(g); \ \Delta H_{sub}$
related to ΔH_{fus} and ΔH_{vap}?

Concept Target

• Understand Hess's law in a context not involving numbers.

Solution

You can think of the sublimation of ice as taking place in two stages. First, the solid melts to liquid, then the liquid vaporizes. The first process has an enthalpy ΔH_{fus}. The second process has an enthalpy ΔH_{vap}. Therefore, the total enthalpy, which is the enthalpy of sublimation, is the sum of these two enthalpies:
   $$\Delta H_{sub} = \Delta H_{fus} + \Delta H_{vap}$$

Conceptual Problem 6.25

A small car is traveling at twice the speed of a larger car, which has twice the mass of the smaller car. Which car has the greater kinetic energy? (Or do they both have the same kinetic energy?)

Concept Target

• Explore the relation of kinetic energy to mass and speed in a nonnumerical context.

Solution

Kinetic energy is proportional to mass and to speed squared. Compare the kinetic energy of the smaller car with that of the larger car (with twice the mass), assuming both are traveling at the same speed. The larger car would have twice the kinetic energy of the smaller car. Or, we could say that the smaller car has only half the kinetic energy of the larger car. Now suppose the speed of the smaller car is increased by a factor of two (so it is now moving at twice its original speed). Its kinetic energy is increased by a factor of four. Therefore, the smaller car now has one-half times four times, or twice, the kinetic energy of the larger car. The smaller car has the greater kinetic energy.

Conceptual Problem 6.26

The equation for the combustion of butane, C_4H_{10}, is
$$C_4H_{10}(g) + (13/2)O_2(g) \rightarrow 4CO_2(g) + 5H_2O(g)$$
Which one of the following generates the least heat? Why?
a. Burning one mole of butane.
b. Reacting one mole of oxygen with excess butane.
c. Producing one mole of carbon dioxide by burning butane.
d. Producing one mole of water by burning butane.

Concept Target

• Explore the relation between heat of reaction and moles of substance in a chemical equation.

Solution

The equation says that 1 mol of butane reacts with (13/2) mol of oxygen to yield 4 mol carbon dioxide and 5 mol water. The reaction yields a certain amount of heat, which you can symbolize as q. So a. yields heat q. On the other hand, b. is only 1 mol oxygen, not (13/2) mol. So, b. yields heat equal to $(2/13)\, q$. This result might be easier to see by first looking at c. Note that the one mole of carbon dioxide stated in c. is only ¼ that given in the equation. This means that c. yields ¼ q (just the inverse of the coefficient in the equation). Similarly, d. yields (1/5) q. Therefore, b. yields the least heat.

Conceptual Problem 6.27

What is the enthalpy change for the preparation of one mole of liquid water from the elements, given the following?

$H_2(g) + \frac{1}{2}O_2(g) \rightarrow H_2O(g)$; ΔH_f

$H_2O(l) \rightarrow H_2O(g)$; ΔH_{vap}

Concept Target

• Understand Hess's law in a context not involving numbers.

Solution

You can imagine this process taking place in two steps: first, the preparation of water vapor from the elements, second, the change of the vapor to liquid. Here are the equations:

$H_2(g) + \frac{1}{2}O_2(g) \rightarrow H_2O(g)$; ΔH_f

$H_2O(g) \rightarrow H_2O(l)$; $-\Delta H_{vap}$

The last equation is the reverse of the vaporization of water, so the enthalpy of the step is the negative of the enthalpy of vaporization. The enthalpy change for the preparation of one mole of liquid water, ΔH, is the sum of the enthalpy changes for these two steps:

$\Delta H = \Delta H_f + (-\Delta H_{vap}) = \Delta H_f - \Delta H_{vap}$

Conceptual Problem 6.28

A block of aluminum and a block of iron, both having the same mass, are removed from a freezer and placed outside on a warm day. When the same quantity of heat has flowed into each block, which block will be warmer? Assume that neither block has yet reached the outside temperature. (See Table 6.1 for the specific heats of the metals.)

Concept Target

• Apply specific heat capacity in a context that doesn't involve calculations.

Solution

The expression for the heat is $q = s \times m \times \Delta t$. For the same amount of heat and mass, the product $s \times \Delta t$ must be constant. The metal with the smaller specific heat will have the larger

Δt. Since the specific heat for aluminum (0.901 J/g°C) is larger than for iron (0.449 J/g°C), the block of iron will have the larger Δt and be warmer.

Conceptual Problem 6.29

You have two samples of different metals, Metal A and metal B, each having the same mass. You heat both metals to 95°C and then place each one into a separate beaker containing the same quantity of water at 25°C.

a. You measure the temperatures of the water in the two beakers when each metal has cooled by 10°C and find that the temperature of the water with metal A is higher than the temperature of the water with metal B. Which metal has the greatest specific heat? Explain.

b. After waiting a period of time, the temperature of the water in each beaker rises to a maximum value. In which beaker does the water rise to the higher value, the one with metal A or metal B? Explain.

Concept Target

• Use experimental data to qualitatively explore heat capacity relationships.

Solution

a. The heat lost by the metal is equal to the heat gained by the water. Since $q = s \times m \times \Delta t$, the heat gained by the water is directly proportional to Δt. Since Δt is larger for metal A, it lost more heat. Now, each metal has the same mass and Δt, so the specific heat is directly proportional to q. Since q is larger for A, the specific heat is larger for A.

b. The metal with the higher specific heat will have absorbed more heat to reach the starting temperature of 95°C, therefore, it will release more heat to the water causing the water to reach a higher temperature. The beaker with metal A will rise to the higher temperature.

Conceptual Problem 6.30

Consider the reactions of silver metal, Ag(s), with each of the halogens: fluorine, $F_2(g)$; chlorine, $Cl_2(g)$; and bromine, $Br_2(l)$. What chapter data could you use to decide which reaction is most exothermic? Which one is?

Concept Target

• Deepen the understanding of the concept of enthalpy of formation.

Solution

Silver metal reacts with a halogen to produce the corresponding silver halide. For example, silver reacts with fluorine to produce silver fluoride. Each reaction corresponds to the reaction forming a silver halide, so you look in the table of enthalpies of formation of compounds (Table 6.2). The most exothermic reaction would be the one with the most negative enthalpy of formation. That would be the reaction for the formation of silver fluoride.

Conceptual Problem 6.31

Tetraphosphorus trisulfide, P_4S_3, burns in excess oxygen to give tetraphosphorus decoxide, P_4O_{10}, and sulfur dioxide, SO_2. Suppose you have measured the enthalpy change for this reaction. How could you use it to obtain the enthalpy of formation of P_4S_3? What other data do you need?

Concept Target

• Explore the relation between enthalpy of a reaction and the enthalpies of formation of substances in the reaction.

Solution

Let us write ΔH_{rxn} for the enthalpy change when one mole of P_4S_3 burns in O_2 to give P_4O_{10} and SO_2. In principle, you could calculate ΔH_{rxn} from enthalpies of formation for the reactants and products of this reaction. You would require the values for P_4S_3, O_2 (which equals zero), P_4O_{10}, and SO_2. Enthalpies of formation for the products, P_4O_{10} and SO_2, are given in Table 6.2 in the text. This means that, if you have measured ΔH_{rxn}, you can use the enthalpies of formation of P_4O_{10} and SO_2 to calculate the enthalpy of formation for P_4S_3. What you have done is this: you have used enthalpies of combustion to calculate enthalpies of formation. This is the idea most often used to obtain enthalpies of formation.

Conceptual Problem 6.32

A soluble salt, MX_2, is added to water in a beaker. The equation for the dissolving of the salt is:

$$MX_2(s) \rightarrow M^{2+}(aq) + 2X^-(aq); \Delta H > 0$$

a. Immediately after the salt dissolves, is the solution warmer or colder?
b. Indicate the direction of heat flow, in or out of the beaker, while the salt dissolves.
c. After the salt dissolves and the water returns to room temperature, what is the value of q for the system?

Concept Target

• Keep track of the direction and magnitude of heat flow.

Solution

a.   Since ΔH is positive, the reaction is endothermic, and the solution will be colder.
b.   While the salt dissolves, heat will flow into the beaker to raise the temperature of the water back to the initial temperature.
c.   After the water returns to room temperature, q for the system will be zero.

Chapter 7

Quantum Theory of the Atom

Concept Check 7.1

Laser light of a specific frequency falls on a crystal that converts this light into one with double the original frequency. How is the wavelength of this frequency-doubled light related to the wavelength of the original laser light? Suppose the original laser light was red. What region of the spectrum would the frequency-doubled light be in? (If this is the visible region, what color is the light?)

Concept Target

• Strengthen understanding of the relation between frequency and wavelength.

Solution

The frequency and wavelength of light are inversely related. Therefore, if the frequency is doubled, the wavelength is halved. Red light has a wavelength around 700 nm, so doubling its frequency halves its wavelength to about 350 nm, which is in the ultraviolet, just beyond the visible spectrum.

Concept Check 7.2

An atom has a line spectrum consisting of a red line and a blue line. Assume that each line corresponds to a transition between two adjacent energy levels. Sketch an energy-level diagram with three energy levels that might explain this line spectrum, indicating the transitions on this diagram. Consider the transition from the highest energy level on this diagram to the lowest energy level. How would you describe the color or region of the spectrum corresponding to this transition?

Concept Target

• Strengthen the qualitative understanding of the relationships among energy levels, transitions, and the regions of the spectrum associated with the transitions.

Solution

Since the transitions are between adjacent levels, the energy-level diagram must look something like the following diagram, with the red transition between two close levels and the blue transition between two levels more widely spaced. (The three levels could be spaced so that the red and blue transitions are interchanged, with the blue transition above the red one.)

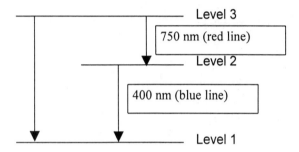

The transition from the top level to the lowest level would correspond to a transition that is greater in energy change than either of the other two transitions. Thus, the three transitions, from lowest to highest energy change, are in the order: red, blue, and the transition from the highest to lowest level. This last transition would have the highest frequency and therefore the shortest wavelength. It would lie just beyond the blue portion of the visible spectrum, in the ultraviolet region.

Concept Check 7.3

A proton is approximately 2000 times heavier than an electron. How would the speeds of these particles compare if their corresponding wavelengths were to be about equal?

Concept Target

• Develop the relationships among quantities in the de Broglie's equation or relation.

Solution

The de Broglie relation says that the wavelength of a particle is inversely proportional to both mass and speed. So, to maintain the wavelength constant while the mass increases would mean that the speed would have to decrease. In going from a particle with the mass of an electron to that of a proton, the speed would have to decrease by a factor of about 2000 in order to maintain the same wavelength. The proton would have to have a speed approximately 2000 times slower than an electron of the same wavelength.

Conceptual Problem 7.19

Consider two beams of the same yellow light. Imagine that one beam has its wavelength doubled; the other has its frequency doubled. Which of these two beams is then in the ultraviolet region?

Concept Target

• Underscore the relations among wavelength, frequency, and region of the spectrum.

Solution

Wavelength and frequency are inversely related. Moreover, ultraviolet light is at higher frequency than yellow light. Doubling the frequency of a beam of light would give that beam a higher frequency than yellow light, whereas doubling the wavelength would give that beam a lower frequency than yellow light. Consequently, the beam with frequency doubled must be the one in the ultraviolet region.

Here is another way to look at the problem. Energy is directly related to the frequency and inversely related to the wavelength. Thus, the beam whose frequency is doubled will increase in energy, whereas the beam whose wavelength is doubled will decrease in energy. Since yellow light is in the visible region of the spectrum, which is lower in energy than the ultraviolet region (Figure 7.5), the beam whose frequency is doubled will be higher in energy and thus in the UV region of the spectrum.

Conceptual Problem 7.20

Some infrared radiation has a wavelength that is 1000 times larger than that of a certain visible light. This visible light has a frequency that is 1000 times smaller than that of some X radiation. How many times more energy is there in a photon of this X radiation than there is in a photon of the infrared radiation?

Concept Target

• Underscore the relationships among wavelength, frequency, and energy of light.

Solution

Frequency is inversely related to the wavelength. Thus, the infrared radiation with a wavelength that is one thousand times larger than the visible light would have a frequency one thousand times smaller than the visible light. But, the visible light has a frequency one thousand times smaller than that of the X radiation. This makes the frequency of the X radiation one million times larger than the frequency of the infrared radiation. Therefore, since energy is directly related to the frequency, the energy of the X radiation would be one million times as large as the energy of the infrared radiation.

Conceptual Problem 7.21

One photon of green light has less than twice the energy of two photons of red light. Consider two hypothetical experiments. In one experiment, potassium metal is exposed to one photon of green light; in another experiment, potassium metal is exposed to two photons of red light. In one of these experiments, no electrons are ejected by the photoelectric effect (no matter how many times this experiment is repeated), whereas in the other experiment at least one electron is ejected. What is the maximum number of electrons that could be ejected during this experiment, one or two?

Concept Target

• Examine the explanation of the photoelectric effect.

Solution

That one color of light does not result in an ejection of electrons implies that that color has too little energy per photon. Of the two colors, red and green, red light has less energy per photon. So you expect that the experiment with red light results in no ejection of photons, whereas the experiment with green light must be the one that ejects electrons. (Two red photons have more than enough energy to eject an electron, but this energy needs to be concentrated in only one photon to be effective.) In the photoelectric effect, one photon of light ejects at most one electron. Therefore, in the experiment with green light, one electron is ejected.

Conceptual Problem 7.22

An atom in its ground state absorbs a photon (photon 1), then quickly emits another photon (photon 2). One of these photons corresponds to ultraviolet radiation, whereas the other one corresponds to red light. Explain what is happening. Which electromagnetic radiation, ultraviolet or red light, is associated with the emitted photon (photon 2)?

Concept Target

• Note the relationship between energy levels and the relative region of the electromagnetic spectrum absorbed or emitted.

Solution

Ultraviolet radiation is higher in energy than red light (Figure 7.5). Since an atom that started in the ground state cannot emit more energy than it absorbed, the absorbed photon must be higher in energy than the emitted photon. This makes the emitted photon (photon 2) the red light.

Conceptual Problem 7.23

The three emission lines involving three energy levels in the magnesium atom occur at wavelengths x, $1.5x$, and $3.0x$ nm. Which wavelength corresponds to the transition from the highest to the lowest of these three energy levels?

Concept Target

• Emphasize the relation between energy levels and wavelength of a transition between energy levels.

Solution

Energy is inversely proportional to the wavelength of the radiation. The transition from the highest to the lowest energy levels would involve the greatest energy change and thus the shortest wavelength, x nm.

Conceptual Problem 7.24

An atom emits yellow light when an electron makes the transition from the $n = 5$ to the $n = 1$ level. In separate experiments, suppose you bombarded the $n = 1$ level of this atom with red light, yellow light (obtained from the previous emission), and blue light. In which experiment or experiments would the electron be promoted to the $n = 5$ level?

Concept Target

• Relate the color of light emission from an atom to the energy level spacing in the atom.

Solution

In a transition from the $n = 1$ to the $n = 5$ energy level, an atom will absorb a photon with the same energy as the photon that was emitted in the transition from the $n = 5$ to the $n = 1$ energy level. Since yellow light was emitted, the experiment using yellow light will promote the electron to the $n = 5$ level.

Conceptual Problem 7.25

Which of the following particles has the longest wavelength?
a.   an electron traveling at x m/s
b.   a proton traveling at x m/s
c.   a proton traveling at $2x$ m/s

Concept Target

• Understand the relationships among the wavelength of a particle, its mass, and its speed.

Solution

A proton is approximately 2000 times the weight of an electron. Also, from the de Broglie relation, $l = h/mv$, you see that the wavelength is inversely proportional to both the mass and the speed of the particle. Considering the protons in parts b. and c., since the mass is the same in both parts, the proton with the smaller speed, part b., will have a longer wavelength. Now, comparing the electron in part a. with the proton in part b., since both have the same speed, the electron in part a. with the smaller mass will have the longer wavelength. Therefore, the electron in part a. will have the longest wavelength.

Conceptual Problem 7.26

Imagine a world in which the rule for the l quantum number is that values start with 1 and go up to n. The rules for the n and m_l quantum numbers are unchanged from those of our world. Write the quantum numbers for the first two shells (i.e., $n = 1$ and $n = 2$).

Concept Target

• Explore the concept of allowable quantum numbers by looking at the quantum numbers for an imaginary world.

Solution

For the first shell, the quantum numbers would have the following allowable values:

$n = 1$; $l = 1$; $m_l = 0, +1, -1$.

For the second shell, the quantum numbers would have the following allowable values:

$n = 2$; $l = 1$; $m_l = 0, +1, -1$

$l = 2$; $m_l = 0, +1, -1, +2, -2$

Chapter 8

Electron Configurations and Periodicity

Concept Check 8.1

Imagine a world in which the Pauli principle is "No more than one electron can occupy an atomic orbital, irrespective of its spin." How many elements would there be in the second row of the periodic table, assuming that nothing else is different about this world?

Concept Target

• Understand how the Pauli principle affects the appearance of the periodic table by looking at a hypothetical situation.

Solution

The second period elements are those in which the $2s$ and $2p$ orbitals fill. Each orbital can hold only one electron, so all four orbitals will be filled after four electrons. Therefore, the second period will have four elements.

Concept Check 8.2

Two elements in Period 3 are adjacent to one another in the periodic table. The ground-state atom of one element has only s electrons in its valence shell; the other one has at least one p electron in its valence shell. Identify the elements.

Concept Target

• Relate the appearance of the periodic table to the electronic structure of the elements.

Solution

The s orbital fills in the first two elements of the period (Groups IA and IIA); then, the p orbital starts to fill (Group IIIA). Thus, the first element is in Group IIA (Mg) and the next element is in Group IIIA (Al).

Concept Check 8.3

Given the following information for element E, identify the element's group in the periodic table: The electron affinity of E is positive (that is, it does not form a stable negative ion). The first ionization energy of E is less than the second ionization energy, which in turn is very much less than its third ionization energy.

Concept Target

• Determine the position of an element in the periodic table from properties of the element.

Solution

From the information given, the element must be in Group IIA. These elements have positive electron affinities and also have large third ionization energies.

Concept Check 8.4

What is the name of the element that is a metalloid with an acidic oxide of formula R_2O_5?

Concept Target

• Relate the nonmetallic-metallic characteristics of an element to the position of the element in the periodic table.

Solution

A metalloid is an element near the staircase line in the periodic table (the green elements in the periodic table on the inside front cover of the book). The formula R_2O_5 suggests a period VA element. There are two metalloids in Group VA, arsenic and antimony. That this is an acidic oxide indicates that this metalloid has considerable nonmetal character. So of the two metalloids, the one nearer the top of the column, arsenic, seems most likely. This is in agreement with the text, which notes that arsenic(V) is acidic, whereas antimony(V) oxide is amphoteric.

Conceptual Problem 8.25

Suppose that the Pauli exclusion principle were "No more than two electrons can have the same four quantum numbers." What would be the electron configurations of the ground states for the first six elements of the periodic table, assuming that, except for the Pauli principle, the usual building-up principle holds?

Concept Target

• Understand how the Pauli principle affects the appearance of the periodic table by looking at a hypothetical situation.

Solution

This statement of the Pauli principle implies that there can be two electrons with the same spin in a given orbital. Because an electron can have either one of two spins, any one orbital can hold a maximum of four electrons. The first six elements of the periodic table would have the following electronic configurations:

(1) $1s^1$
(2) $1s^2$
(3) $1s^3$
(4) $1s^4$
(5) $1s^4 2s^1$
(6) $1s^4 2s^2$

Conceptual Problem 8.26

Imagine a world in which all quantum numbers, except the l quantum number, are as they are in the real world. In this imaginary world, l begins with 1 and goes up to n (the value of the principal quantum number). Assume that the orbitals fill in the order by n, then l; that is, the first orbital to fill is for $n = 1$, $l = 1$; the next orbital to fill is for $n = 2$, $l = 1$, and so forth. How many elements would there be in the first period of the periodic table?

Concept Target

• Understand how the rules for quantum numbers affect the periodic table by looking at an imaginary world.

Solution

The first period of the periodic table would have the following allowed quantum numbers: $n = 1$; $l = 1$; $m_l = 0, +1, -1$; $m_s = +1/2, -1/2$. There are six different possible combinations. Therefore, there would be six elements in the first period.

Conceptual Problem 8.27

Two elements in Period 5 are adjacent to one another in the periodic table. The ground-state atom of one element has only *s* electrons in its valence shell; the other has at least one *d* electron in an unfilled shell. Identify the elements.

Concept Target

• Use the structure of the periodic table to identify elements of given electronic structure.

Solution

The elements are in Group IIA (only *s* electrons) and IIIB (*d* electrons). They are also in Period 5. Therefore, the elements are strontium (Sr) and yttrium (Y).

Conceptual Problem 8.28

Two elements are in the same column of the periodic table, one above the other. The ground-state atom of one element has two *s* electrons in its outer shell, and no *d* electrons anywhere in its configuration. The other element has *d* electrons in its configuration. Identify the elements.

Concept Target

• Use the structure of the periodic table to identify elements of given electronic structure.

Solution

The elements are in Group IIA. They must also be in Periods 4 (no *d* electrons) and 5 (*d* electrons). Therefore, the elements are calcium (Ca) and strontium (Sr).

Conceptual Problem 8.29

You travel to an alternate universe where the atomic orbitals are different from those on earth, but all other aspects of the atoms are the same. In this universe, you find that the first (lowest energy) orbital is filled with three electrons and the second orbital can hold a maximum of nine electrons. You discover an element *Z* that has five electrons in its atom. Would you expect *Z* to be more likely to form a cation or an anion? Indicate a possible charge on this ion.

Concept Target

• Adapt the rules of orbital filling to a new set of conditions.

Solution

Keeping in mind that a filled orbital is usually a stable configuration for an atom, an element in this universe with five electrons would probably lose the two electrons in the second orbital and form a cation with a charge of positive two. The other possible option is for the atom to gain seven additional electrons to fill the second orbital; however, this is unlikely given that the nuclear charge would be relatively small and electron-electron repulsions in such an atom would be large.

Conceptual Problem 8.30

Would you expect to find an element having both a very large (positive) first ionization energy and an electron affinity that is much less than zero (large but negative)? Explain.

Concept Target

• Explore the relationship between ionization energy and electron affinity.

Solution

Keep in mind that the ionization energy of an atom provides a measure of how strongly an electron is attracted to that atom. The electron affinity of an atom provides a measure of how strongly attracted an additional electron is to the atom. Both electron affinity and ionization energy provide information about the strength of the attraction between electrons and a particular nucleus. An element that forms an anion easily has an electron affinity much less than zero, and a very large first ionization energy. Examples are the elements on the upper right of the periodic table, like fluorine, with an electron affinity of -328 kJ/mol and a first ionization energy of 1681 kJ/mol.

Conceptual Problem 8.31

Two elements are in the same group, one following the other. One is a metalloid, the other is a metal. Both form oxides of the formula RO_2; the first is acidic, the next is amphoteric. Identify the two elements.

Concept Target

• Relate the nonmetallic-metallic characteristics of elements to their position in the periodic table and their identity.

Solution

The elements that form oxides of the form RO_2 are in Groups IVA and VIA. However, there are no metals in Group VIA. Therefore, the elements are in Group IVA. The metalloid is germanium (Ge), and the metal is tin (Sn). GeO_2 is the acidic oxide, and SnO_2 is the amphoteric oxide.

Conceptual Problem 8.32

A metalloid has an acidic oxide of the formula R_2O_3. The element has no oxide of the formula R_2O_5. What is the name of the element?

Concept Target

• Relate the acid-base character and formula of the oxide of an element to the position of the element in the periodic table, from which you can identify the element.

Solution

Oxides of Groups IIIA and VA have oxides of the form R_2O_3. However, Group VA oxides can also be of the form R_2O_5, so the element is in group IIIA. It is also acidic. Therefore, it must be boron oxide, B_2O_3.

Chapter 9

Ionic and Covalent Bonding

Concept Check 9.1

The following are electron configurations of some ions. Which ones would you expect to see in chemical compounds? State the concept or rule you used to decide for or against any ion.
a. Fe^{2+} $[Ar]3d^4 4s^2$
b. N^{2-} $[He]2s^2 2p^5$
c. Zn^{2+} $[Ar]3d^{10}$
d. Na^{2+} $[He]2s^2 2p^6$
e. Ca^{2+} $[Ne]3s^2 3p^6$

Concept Target

• Use concepts to decide for or against an electron configuration of an ion you would expect to see in a compound.

Solution

a. The 2+ ions are common in transition elements, but it is the outer s electrons that are lost to form these ions in compounds. Iron, whose configuration is $[Ar]3d^6 4s^2$, would be expected to lose the two $4s$ electrons to give the configuration $[Ar]3d^6$ for the Fe^{2+} ion in compounds. The configuration given in the problem is for an excited state; you would not expect to see it in compounds.

b. Nitrogen, whose ground-state atomic configuration is $[He]2s^2 2p^3$, would be expected to form an anion with a noble-gas configuration by gaining three electrons. This would give the anion N^{3-} with the configuration $[He]2s^2 2p^6$. You would not expect to see the anion N^{2-} in compounds.

c. The zinc atom has the ground-state configuration $[Ar]3d^{10} 4s^2$. The element is often considered to be a transition element. In any case, you expect the atom to form compounds by losing its $4s$ electrons to give Zn^{2+} with the pseudo-noble-gas configuration $[Ar]3d^{10}$. This is the ion configuration given in the problem.

d. The configuration of the ground-state sodium atom is $[He]2s^22p^53s^1$. You would expect the atom to lose one electron to give the Na^+ ion with the noble-gas configuration $[He]2s^22p^6$. You would not expect to see compounds with the Na^{2+} ion.

e. The ground state of the calcium atom is $[Ne]3s^23p^64s^2$. You would expect the atom to lose its two outer electrons to give the Ca^{2+} ion with noble-gas configuration $[Ne]3s^23p^6$, which is the configuration given in the problem.

Concept Check 9.2

Each of the following may seem, at first glance, to be plausible electron-dot formulas for the molecule N_2F_2. Most, however, are incorrect for some reason. What concepts or rules apply to each, either to cast it aside or to keep it as the correct formula?

a. $:\ddot{F}:\ddot{N}:\ddot{N}:\ddot{F}:$

b. $:\ddot{F}:\ddot{N}::\ddot{N}:\ddot{F}:$

c. $:\ddot{F}::\ddot{N}:\ddot{N}:\ddot{F}:$

d. $:\ddot{F}:\ddot{N}:\ddot{N}:\ddot{F}:$

e. $:\ddot{F}:\ddot{N}::\ddot{F}:\ddot{N}:$

f. $:\ddot{F}:\ddot{N}\ \ \ddot{N}:\ddot{F}:$

Concept Target

• Apply concepts or rules to decide on the appropriateness of an electron-dot formula.

Solution

a. There are two basic points to consider in assessing the validity of each of the formulas given in the problem. One is whether the formula has the correct skeleton structure. You expect the F atoms to be bonded to the central N atoms because the F atoms are more electronegative. The second point is the number of dots in the formula. This should equal the total number of electrons in the valence shells of the atoms (5 for each nitrogen atom and 7 for each fluorine atom), which is $(2 \times 5) + (2 \times 7) = 24$, or 12 pairs. The number showing in the formula here is 13, which is incorrect.

b. This formula has the correct skeleton structure and the correct number of dots. All of the atoms have octets, so the formula would appear to be correct. As a final check, however, you might try drawing the formula beginning with the skeleton structure. In drawing an electron-dot formula, after connecting atoms by single bonds (a single electron pair) you would place electron pairs around the outer atoms (the F atoms in this formula) to give octets. After doing that, you would have used up 9 electron pairs (3 for the single bonds and 3 for each F atom to fill out its octet). This leaves 3 pairs, which you might distribute as follows:

$$: \overset{..}{\underset{..}{F}} : \overset{..}{N} : \overset{..}{\underset{..}{N}} : \overset{..}{\underset{..}{F}} :$$

One of the nitrogen atoms (the one on the right) does not have an octet. The lack of an octet on this atom suggests that you try for a double bond. This suggests that you move one of the lone pairs on the left N atom into one of the adjacent bonding regions. Moving this lone pair into the N—N region would give a symmetrical result, whereas moving the lone pair into the F—N region would not. The text notes that the atoms often showing multiple bonds are C, N, O, and S. The formula given here with a nitrogen-nitrogen double bond appears quite reasonable.

c. This formula is similar to the previous one, b., but the double bond is between the F and N atoms. Multiple bonds to F are less likely than those between two N atoms. So this is not the preferred formula. You could also apply the rules of formal charge in this case, and you would come to the same conclusion. The formula here gives a -1 charge to the left F atom and a +1 charge to the right N atom, whereas each of the atoms in the b. formula has zero formal charges. Rule A says that whenever you can write several Lewis formulas for a molecule, choose the formula having the lowest magnitudes of formal charges. In this case, this is the formula b. Strictly speaking, both b. and c. could be regarded as resonance formulas, but b. would have much more importance than c. in describing the electronic structure of the molecule.

d. This formula is similar to the one you drew earlier in describing how you would get to the formula in b. The left N atom does not have an octet, which suggests you move a lone pair on the other N atom into the N—N bond region to give a double bond.

e. This formula does not have the correct skeleton structure.

f. This formula has the correct skeleton structure and the correct total number of electron dots, but neither N atom has an octet. In fact, there is no bond between the two N atoms.

Concept Check 9.3

Which of the models shown below most accurately represents the hydrogen cyanide molecule, HCN? Write the electron-dot formula that most closely agrees with this model. State any concept or rule you used in arriving at your answer.

(a)   (b)   (c)   (d)

Concept Target

- Use VSEPR and formal charge concepts to determine the correct molecular structure of a molecule.
- Write the correct Lewis formula of a molecule from its molecular model.

Solution

a. This model and corresponding Lewis structure, H:C:::N:, has the expected skeleton structure. (H must be an exterior atom, but either C or N might be in the center; however, you expect the more electropositive, or less electronegative atom, C, to be in this position.) This formula also has the correct number of electron dots ($1 + 4 + 5 = 10$, or 5 pairs). Finally, the formal charge of each atom is zero. Therefore, this model should be an accurate representation of the HCN molecule.

b. This structure has the more electronegative atom, N, in the central position; you don't expect this to be the correct structure. You can also look at this from the point of view of formal charges. This formula has a +1 charge on the N atom and a -1 charge on the C atom. You would not expect the more electronegative atom to have the positive formal charge. Moreover, the previous formula, a., has zero charges for each atom, which would be preferred.

c. If you draw the Lewis structure, you will see that each atom has an octet, but the formula has too many electron pairs (7 instead of 5). You can remove two pairs and still retain octets if you move two lone pairs into the bonding region, giving a triple bond.

d. If you draw the Lewis structure you will find the H atom has a bond and a lone pair attached to it, so this is not a reasonable structure. You would arrive at the same conclusion using formal charges. The formal charges are -2 for H, +1 for C, and +1 for N. Since you already know that formula a. has zero formal charges for each atom, such high formal charges for the atoms in formula d. mean that it is not a very good representation of the molecule.

Conceptual Problem 9.19

You land on a distant planet in another universe and find that the $n = 1$ level can hold a maximum of 4 electrons, the $n = 2$ level can hold a maximum of 5 electrons, and the $n = 3$ level can hold a maximum of 3 electrons. Like our universe, protons have a charge of $+1$, electrons have a charge of -1, and opposite charges attract. Also, a filled shell results in greater stability of an atom, so the atom tends to gain or lose electrons to give a filled shell. Predict the formula of a compound that results from the reaction of a neutral metal atom X, which has 7 electrons, and a neutral nonmetal atom Y, which has three electrons.

Concept Target

• Apply the rules of orbital filling and the octet rule to a different set of circumstances.

Solution

Because the compound that forms is a combination of a metal and a nonmetal, we would expect it to be ionic. If we assume that metal atoms tend to lose electrons to obtain filled shells, then the metal atom X would lose three electrons from the $n = 2$ level forming the X^{3+} cation. We can expect the nonmetal atom Y to gain electrons to obtain a filled shell, so it requires an additional electron to fill the $n = 1$ level forming the Y^- anion. To produce an ionic compound with an overall charge of zero, the compound formed from these two elements would be XY_3.

Conceptual Problem 9.20

Which of the following represent configurations of thallium ions in compounds? Explain your decision in each case.
a. Tl^{2+} $[Xe]4f^{14}5d^{10}6p^1$
b. Tl^{3+} $[Xe]4f^{14}5d^{10}$
c. Tl^{4+} $[Xe]4f^{14}5d^9$
d. Tl^+ $[Xe]4f^{14}5d^{10}6s^2$

Concept Target

• Emphasize the configurations expected for ions of main-group elements.

Solution

Elements in higher periods in Group IIIA tend to form ions by losing either the outer p electron to form $+1$ ions or by losing the outer s and p electrons to form $+3$ ions. In the case of thallium, these ions have the configurations in d. and b., respectively. The $+2$ ion given in a. represents a loss of the two outer s electrons, leaving the outer p electron of higher energy.

This configuration would represent an excited state of the +2 ion. You would not expect a +2 ion in thallium compounds, and an excited configuration is even less likely. The +4 ion given in c. is formed by loss of the outer s and p electrons plus a d electron from the closed d subshell. This would be an unlikely ion for thallium compounds.

Conceptual Problem 9.21

Examine each of the following electron-dot formulas and decide whether the formula is correct, or whether you could write a formula that better approximates the electron structure of the molecule. State which concepts or rules you use in each case to arrive at your conclusion.

a. $:\overset{..}{\underset{}{N}}:\overset{..}{\underset{}{N}}:$

b. $:\overset{..}{\underset{..}{Cl}}:C:::N:$

c. $H:C::\overset{..}{\underset{..}{F}}$
    $\quad :\overset{}{\underset{..}{O}}:$

d. $\quad\quad :\overset{..}{\underset{..}{O}}:$
    $H:\overset{..}{\underset{..}{O}}::\overset{..}{\underset{..}{C}}::\overset{..}{\underset{..}{O}}:H$

Concept Target

• Use concepts and rules to decide on the correctness of an electron-dot formula.

Solution

a. Incorrect. The atoms in this formula do not obey the octet rule. The formula has the correct number of valence electrons, so this suggests a multiple bond between the N atoms.

b. Correct. The central atom is surrounded by more electronegative atoms, as you would expect, and each atom obeys the octet rule.

c. Incorrect. The skeleton structure is acceptable (the central atom is surrounded by more electronegative atoms), but you would expect the double bond to be between C and O, rather than C and F (C, N, O, and S form multiple bonds). You would come to this same conclusion using rules of formal charge (the formula has a formal charge of +1 on F and -1 on O, whereas you would expect these formal charges to be interchanged, with the negative charge on the F atom which is more electronegative).

d. Incorrect. The skeleton structure is OK, but the carbon atom has 10 valence electrons about it. This suggests that you replace the two carbon-oxygen double bonds by one carbon-oxygen double bond (because there is one extra pair of electrons on the C atom). The most symmetrical location of the double bond uses the oxygen atom not bonded to an H atom. Also, only this formula has zero formal charges on all atoms.

Conceptual Problem 9.22

For each of the following molecular models, write the appropriate Lewis formula.

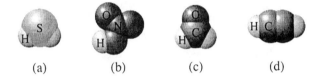

(a)   (b)   (c)   (d)

Concept Target

• Write Lewis formulas from molecular models.

Solution

Be sure that all atoms on your structure have the correct number of valence electrons.

a. H–S̈–H   b. H–Ö–N–Ö̈   c. H–C̈–H

d. H–C≡C–H

Conceptual Problem 9.23

For each of the following molecular formulas, draw the most reasonable skeleton structure.
a. CH_2Cl_2
b. HNO_2
c. NOF
d. N_2O_4

Concept Target

• Use concepts and rules to decide on the skeleton structure of a molecule.

Solution

a. To arrive at a skeleton structure, you decide which is the central atom (it cannot be H). The C atom is less electronegative than the Cl atom, so you place it as the central atom and surround it by the other atoms.

$$
\begin{array}{c}
\text{Cl} \\
| \\
\text{H—C—Cl} \\
| \\
\text{H}
\end{array}
$$

b. HNO_2 is an oxyacid in which O atoms bond to the central atom, with the H bonded to O. The central atom must be N (the only other atom), so the skeleton structure is
    H—O—N—O

c. You place the least electronegative atom (N) as the central atom and bond it to the other atoms.
    F—N—O

d. N is less electronegative than O. The most symmetrical structure would have the two N atoms in the center with two O atoms bonded to each N atom.

$$
\begin{array}{cc}
\text{O} & \text{O} \\
\diagdown & \diagup \\
\text{N—N} & \\
\diagup & \diagdown \\
\text{O} & \text{O}
\end{array}
$$

Conceptual Problem 9.24

Below are a series of resonance formulas for N_2O (nitrous oxide). Rank these in terms of how closely you think each one represents the true electron structure of the molecule. State the rules and concepts you use to do this ranking.

a. :N̈–N≡O:

b. :N≡N–Ö:

c. :N̈=N=Ö

Concept Target

• Use concepts and rules to decide on the best resonance formula for a molecule.

Solution

The ranking is a., c., b. (best) for the following reasons:

a. This formula has a -2 formal charge on the outer N and +1 charges on the other two atoms. This formula has a larger magnitude of formal charges than either of the other two formulas, so it ranks last as a representation of the electron structure of the N_2O molecule.
b. This formula has a -1 formal charge on O and a +1 formal charge on the central N. This places the negative formal charge on the more electronegative atom, O, making this the best representation of the electron structure of the molecule.
c. This formula has -1 formal charge on the outer N atom and +1 formal charge on the central N atom. This is better than resonance formula a., but not as good as b.

Conceptual Problem 9.25

Sodium, Na, reacts with element X to form an ionic compound with the formula Na_3X.
a. What is the formula of the compound you expect to form when calcium, Ca, reacts with element X?
b. Would you expect this compound to be ionic or molecular?

Concept Target

• Determine the chemical behavior of an unknown element from information about how it reacts with known elements.

Solution

a. In order to be a neutral ionic compound, element X must have a charge of –3. In an ionic compound, calcium forms the Ca^{2+} cation. The combination of Ca^{2+} and X^{3-} would form the ionic compound with the formula Ca_3X_2.
b. Since element X formed an ionic compound with sodium metal, it indicates that it is probably a nonmetal with a high electron affinity. When a nonmetal with a high electron affinity combines with a metal such as calcium, an ionic compound is formed.

Conceptual Problem 9.26

The enthalpy change for each of the following reactions was calculated using bond energies. The bond energies of X-O, Y-O, and Z-O are all equal.

X-X + O=O → X-O-O-X; ΔH = -275kJ
Y-Y + O=O → Y-O-O-Y; ΔH = +275kJ
Z-Z + O=O → Z-O-O-Z; ΔH = -100kJ

a. Rank the bonds X-X, Y-Y, and Z-Z from strongest to weakest.
b. Compare the energies required to completely dissociate each of the products to atoms.
c. If O_2 molecules were O-O instead of O=O, how would this change ΔH for each reaction?

Concept Target

• Explore how bond energy relates to ΔH of reactions.

Solution

a. In general, the enthalpy of a reaction is (approximately) equal to the sum of the bond energies for the bonds broken minus the sum of the bond energies for bonds formed. Therefore, the bond energy (BE) for each of the reactions can be calculated by the following relationships.

$\Delta H = BE(X\text{-}X) + BE(O\text{=}O) - 2 \times BE(X\text{-}O) - BE(O\text{-}O)$
$\Delta H = BE(Y\text{-}Y) + BE(O\text{=}O) - 2 \times BE(Y\text{-}O) - BE(O\text{-}O)$
$\Delta H = BE(Z\text{-}Z) + BE(O\text{=}O) - 2 \times BE(Z\text{-}O) - BE(O\text{-}O)$

Using information that the bond energies of X-O, Y-O, and Z-O are all equal, the difference in the values of ΔH for each reaction is solely a function of the X-X, Y-Y, or Z-Z bonds. Therefore, the compound with the most positive value of ΔH has the strongest bonds. The strongest to weakest bond ranking is Y-Y > Z-Z > X-X.

b. Since the bond strengths of each product are equal, the same amount of energy would be required to dissociate each of the products into atoms.

c. In this case, less energy would be required to break the bonds, so the ΔH for each reaction would decrease.

Chapter 10

Molecular Geometry and Chemical Bonding Theory

Concept Check 10.1

An atom in a molecule is surrounded by four pairs of electrons, one lone pair and three bonding pairs. Describe how the four electron pairs are arranged about the atom. How are any three of these pairs arranged in space? What is the geometry of this atom plus the three bonded to it?

Concept Target

- Apply VSEPR concepts to find the arrangement of electron pairs about an atom and the geometry of a molecule.

Solution

The VSEPR model predicts that four electron pairs about any atom in a molecule will distribute themselves to give a tetrahedral arrangement. Any three of these electron pairs would have a trigonal pyramidal arrangement. The geometry of a molecule having a central atom with three atoms bonded to it would be trigonal pyramidal.

Concept Check 10.2

Two molecules, each with the general formula AX_3, have different dipole moments. Molecule Y has a dipole moment of zero, whereas molecule Z has a nonzero dipole moment. From this information, what can you say about the geometries of Y and Z?

Concept Target

• Use general concepts to deduce information about molecular geometry from the dipole
moment.

Solution

A molecule AX_3 could have one of three geometries: it could be trigonal planar, trigonal
pyramidal, or T-shaped. Assuming that the three groups attached to the central atom are
alike, as indicated by the formula, the planar geometry should be symmetrical, so that even if
the A—X bonds are polar, their polarities would cancel to give a nonpolar molecule (dipole
moment of zero). This would not be the case in the trigonal pyramidal geometry. In this
situation the bonds all point to one side of the molecule. It is possible for such a molecule to
have a lone pair which would point away from the bonds, whose polarity might fortuitously
cancel the bond polarities, although an exact cancellation is not likely (see Figure 10.20). In
general, you should expect the trigonal pyramidal molecule to have a nonzero dipole
moment, but a zero dipole is possible. The argument for the T-shaped geometry is similar to
that for the trigonal pyramidal geometry. The bonds point in a plane, but toward one side of
the molecule. Unless the sum of the bond polarities was fortuitously canceled by polarities
from lone pairs, this geometry would have a nonzero dipole moment. This means that
molecule Y is likely to be trigonal planar, but trigonal pyramidal or T-shaped geometries are
possible. Molecule Z cannot have a trigonal planar geometry, but must be either trigonal
pyramidal or T-shaped.

Concept Check 10.3

An atom in a molecule has one single bond and one triple bond to other atoms. What hybrid
orbitals do you expect for this atom? Describe how you arrived at this result.

Concept Target

• Use the bonding situation in a molecule to decide the hybrid orbitals involved.

Solution

Assuming that there are no lone pairs, the atom has four electron pairs and, therefore, an
octet of electrons about it. The single bond and the triple bond each require a sigma bond
orbital, for a total of two such orbitals. This suggests *sp* hybrids on the central atom.

Conceptual Problem 10.17

Match the following molecular substances with one of the molecular models (i) to (iv) that correctly depicts the geometry of the corresponding molecule.
a. SeO_2 b. $BeCl_2$ c. PBr_3 d. BCl_3

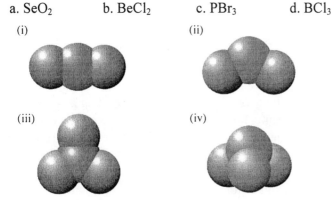

(i)

(ii)

(iii)

(iv)

Concept Target

• Use VSEPR concepts to correctly match molecular formulas with molecular models.

Solution

In order to solve this problem, draw the Lewis structure for each of the listed molecules. In each case, use your Lewis structure to determine the geometry and match this geometry with the correct model.
a. SeO_2 is angular and has an AB_2 geometry. This is represented by model (ii).
b. $BeCl_2$ is linear and has an AB_2 geometry. This goes with model (i).
c. PBr_3 is trigonal pyramidal and has an AB_3 geometry. This goes with model (iv).
d. BCl_3 is trigonal planar and has an AB_3 geometry. This is represented by model (iii).

Conceptual Problem 10.18

Suppose that an ethane molecule, CH_3CH_3, is broken into two CH_3 molecules in such a way that one :CH_3 molecule retains the electron pair that was originally the one making up the C—C bond. The other CH_3 molecule has two fewer electrons. Imagine that momentarily this CH_3 molecule has the geometry it had in the ethane molecule. Describe the electron repulsions present in this molecule and how they would be expected to rearrange its geometry.

Concept Target

• Apply VSEPR concepts to visualize how the repulsions of electron pairs affect the geometry of a molecule.

Solution

In a CH_3CH_3 molecule, each C atom has four electron pairs arranged tetrahedrally. Within this molecule, each CH_3 considered as a separate group has a trigonal pyramidal geometry (with three C—H bonding pairs and a fourth pair from the C—C bond around the C atom). The :CH_3 molecule retains this trigonal pyramidal geometry, having three bonding pairs and one lone pair around the C atom. The CH_3 molecule, however, has only three electron pairs around the C atom. Initially, as the CH_3 molecule breaks away from the ethane molecule, it has the trigonal pyramidal geometry it had in the ethane molecule. However, the repulsions of the bonding electron pairs on the CH_3 molecule are no longer balanced by the fourth pair (from the C—C bond), so the molecule flattens out to form a trigonal planar geometry.

Conceptual Problem 10.19

What are the bond angles predicted by the VSEPR model about the carbon atom in the formate ion, HCO_2^-? Considering that the bonds to this atom are not identical, would you expect the experimental values to agree precisely with the VSEPR values? How might they differ?

Concept Target

• Note how slight differences in sizes of bonding pairs affect bond angles predicted by the VSEPR model.

Solution

The formate ion, HCO_2^-, is expected to have trigonal planar geometry by the VSEPR model. (Each resonance formula has one C=O bond, one C—O bond, and one C—H bond, giving a total of three groups about the C atom.) The VSEPR model predicts bond angles of 120°. However, a bond between the carbon atom and an oxygen atom has a bond order of 3/2 (resonance between a single and a double bond) and requires more room than a pure single bond. The repulsion between the two C—O bonds would be greater than the repulsion between a C—O bond and the C—H bond. Thus, you would predict that the O—C—O angle is slightly greater than 120°, whereas an O—C—H angle is slightly less than 120°.

Conceptual Problem 10.20

An atom in a molecule has two bonds to other atoms and one lone pair. What kind of hybrid orbitals do you expect for this atom? Describe how you arrived at your answer.

Concept Target

• Deduce the type of hybrid orbitals to use for an atom in a molecule.

Solution

The arrangement of electron pairs about this atom suggested by two bonds and one lone pair is trigonal planar. You would expect sp^2 hybrid orbitals for this atom (a total of three hybrid orbitals). Two of these hybrid orbitals would be used to form the two bonds; the third hybrid orbital would be used for the lone pair.

Conceptual Problem 10.21

Two compounds have the same molecular formula, $C_2H_2Br_2$. One has a dipole moment; the other does not. Both compounds react with bromine, Br_2, to produce the same compound. This reaction is a generally accepted test for double bonds, and each bromine atom of Br_2 adds to a different atom of the double bond. What is the identity of the original compounds? Describe the argument you use.

Concept Target

• Underscore the possibility of *cis-trans* isomers in compounds with carbon-carbon double bonds.

Solution

The reaction with Br_2 indicates that $C_2H_2Br_2$ has a double bond. There are three possible isomers of $C_2H_2Br_2$ having double bonds:

Compounds I and III have dipole moments. The addition of Br_2, with one Br going to each C atom, yields the following products:

$$\begin{matrix} Br & Br \\ | & | \\ Br-C-C-Br \\ | & | \\ H & H \end{matrix}$$

IA

$$\begin{matrix} Br & Br \\ | & | \\ Br-C-C-Br \\ | & | \\ H & H \end{matrix}$$

IIA

$$\begin{matrix} Br & H \\ | & | \\ Br-C-C-Br \\ | & | \\ Br & H \end{matrix}$$

IIIA

Products IA and IIA are identical and arise from compounds I and II. Thus, the original two compounds, one not having a dipole moment, the other having a dipole moment, but both reacting with bromine to give the same product are compounds I and II, respectively.

Conceptual Problem 10.22

A neutral molecule is identified as a tetrafluoride, XF_4, where X is an unknown atom. If the molecule has a dipole moment of 0.63 D, can you give some possibilities for the identity of X?

Concept Target

• Practice using a number of bonding concepts and the concept of dipole moment.

Solution

A neutral molecule of the form XF_4 could have either four, five, or six electron pairs around X. With four bonding pairs and no lone pairs, the geometry is tetrahedral. The molecule would be symmetric and nonpolar. Similarly, with four bonding pairs and two lone pairs, the geometry is square planar; the molecule would again be symmetric and nonpolar. However, with four bonding pairs and one lone pair, the geometry would be seesaw; the molecule would be nonsymmetric in shape and could be polar. This fits the description of a compound with a dipole moment of 0.63 D. To identify X, let's look at the total number of valence electrons in XF_4. Each F atom, with its bonding electrons, has an octet (8) of electrons. In addition, there is a lone pair, or 2 electrons. Thus, the total number of valence electrons is 4 x 8 + 2 = 34. Of these, 4 x 7 = 28 are from the F atoms, leaving 34 - 28 = 6 valence electrons coming from X. So X is a VIA element. X cannot be oxygen because a seesaw geometry would require sp^3d hybrid orbitals and oxygen does not have d orbitals to hybridize. Thus, X must be either sulfur (S), selenium (Se), or tellurium (Te).

Chapter 11

States of Matter; Liquids and Solids

Concept Check 11.1

Shown here is a representation of a closed container in which you have just placed 10 L of H_2O. In our experiment, we are going to call this starting point in time $t = 0$ and assume that all of the H_2O is in the liquid phase. We have represented a few of the H_2O molecules in the water as dots.

$t = 0$

a. Consider a time, $t = 1$, at which some time has passed but the system has not reached equilibrium.
   (i)   How will the level of the liquid H_2O compare to that at $t = 0$?
   (ii) How will the vapor pressure in the flask compare to that at $t = 0$?
   (iii) How will the number of H_2O molecules in the vapor state compare to that at $t = 0$?
   (iv) How does the rate of evaporation in this system compare to the rate of condensation?
   (v)   Draw a picture of the system at $t = 1$.

b. Consider a time, $t = 2$, at which enough time has passed for the system to reach equilibrium.
   (i)   How will the level of the liquid H_2O compare to that at $t = 1$?
   (ii) How will the vapor pressure in the flask compare to that at $t = 1$?
   (iii) How will the number of H_2O molecules in the vapor state compare to that at $t = 1$?
   (iv) How does the rate of evaporation in this system compare to the rate of condensation?
   (v)   Draw a picture of the system at $t = 2$.

Concept Target

• Use the chemical concepts of equilibrium, rate, and vapor pressure, to describe the process of evaporation and condensation in a closed container.

Solution

a. (i) At $t = 0$, since the system is not at equilibrium and there are no H_2O molecules in the gaseous state, you would expect the rate of evaporation to exceed the rate of condensation. At $t = 1$, since evaporation has proceeded at a greater rate than condensation, there must now be fewer molecules in the liquid state resulting in a lower level of H_2O (*l*).

(ii) At $t = 1$, since some of the H_2O has gone into the vapor state, the vapor pressure must be higher.

(iii) At $t = 1$, since evaporation has occurred, there must be more molecules in the vapor state.

(iv) At $t = 0$, since the system is not at equilibrium and there are no H_2O molecules in the gaseous state, you would expect the rate of evaporation to exceed the rate of condensation.

(v)

$t = 1$

b. (i) Between $t = 1$ and $t = 2$, the system is still prior to reaching equilibrium. Therefore, the rate of evaporation continues to exceed the rate of condensation so you would expect the water level to be lower.

(ii) Prior to reaching equilibrium at $t = 2$, you would continue to observe a rate of evaporation greater than the rate of condensation resulting in a higher vapor pressure than $t = 1$.

(iii) Since evaporation has been occurring at a greater rate than condensation between points $t = 1$ and $t = 2$, you would expect more molecules in the vapor state at $t = 2$.

(iv) When the system has reached equilibrium at $t = 2$, the rate of evaporation equals the rate of condensation.

(v)

$t = 2$

Concept Check 11.2

When camping at high altitude, you need to pay particular attention to changes in cooking times for foods that are boiled in water. If you like eggs that are boiled for 10 minutes near sea level, would you have to cook them for a longer or shorter time at 3200 m to get the egg you like? Be sure to explain your answer.

Concept Target

• Ensure a grasp of the relationship between the boiling point of a liquid and pressure above the liquid.

Solution

You would have to cook the egg for a longer time. The reason is that since there is lower atmospheric pressure at high altitude, water boils at a lower temperature than near sea level. Since the temperature is lower, it would take longer to transfer an equivalent amount of heat to the egg.

Concept Check 11.3

A common misconception is that the following chemical reaction occurs when boiling water: $2H_2O(l) \rightarrow 2H_2(g) + O_2(g)$ instead of $H_2O(l) \rightarrow H_2O(g)$.
a. What physical evidence do you have that the second reaction is correct?
b. How would the enthalpy of the wrong reaction compare with that of the correct reaction?
c. How could you calculate the enthalpy change for the wrong reaction (see Chapter 6)?

Concept Target

• Address the common misconception that when water boils it produces H_2 (g) and O_2 (g).

Solution

a. If the first reaction occurred, the mixture of hydrogen and oxygen that resulted would form an explosive mixture.
b. Since you would be breaking strong chemical bonds and forming relatively weak bonds, the enthalpy for the first reaction (the wrong reaction) would be many times greater (more positive) than for the second reaction.
c. Apply Hess's law. The enthalpy for the wrong reaction would be equal to two times ΔH_f° for $H_2O(l)$, plus the heat required to raise the temperature of two moles of water from 25°C to 100°C.

Concept Check 11.4

Shown here is a representation of a unit cell for a crystal. The large balls (they are orange in the text) are atom A, and the small black balls are atom B.

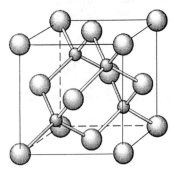

a. What is the chemical formula of the compound that has this unit cell ($A_?B_?$)?
b. Consider the configuration of the A atoms. Is this a cubic unit cell? If so, which type?

Concept Target

• Use the drawing of a unit cell to determine the chemical formula and the structure of a compound.

Solution

a. First, consider the B balls (small). There are four atoms, each completely inside the cell. Thus, there are four B atoms per cell. Next, there are 14 A atoms (large). Of these, eight are in corners and contribute 1/8 to the cell. Six atoms are in faces and contribute 1/2 to the cell. Thus, there are 8 x (1/8) + 6 x (1/2) = 4 A atoms per cell. The ratio of A atoms to B atoms is 4 to 4, or 1 to 1. Thus the formula of the compound is AB.

b. Since all of the B atoms are completely within the cell, the shape of the cell is determined by the A atoms only. It is a face-centered cubic unit cell.

Conceptual Problem 11.21

Shown here is a curve of the distribution of kinetic energies of the molecules in a liquid at an arbitrary temperature T.

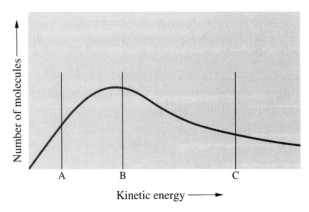

The lines marked A, B, and C represent the point where each of the molecules for three different liquids (liquid A, liquid B, and liquid C) has enough kinetic energy to escape into the gas phase (see Figure 11.5 for more information). Write a brief explanation for each of your answers to the following questions.

a. Which of the molecules – A, B, or C – would have the majority of the molecules in the gas phase at temperature T?
b. Which of the molecules – A, B, or C – has the strongest intermolecular attractions?
c. Which of the molecules would have the lowest vapor pressure at temperature T?

Concept Target

• Relate the information presented in an energy distribution curve to the observed physical properties of liquids and gases.

Solution

a. The number of molecules in the gas phase is directly related to the kinetic energy needed to escape into the gas phase. The molecules with the most kinetic energy will have the most molecules in the gas phase. Since molecules of C have the highest kinetic energy, they will have the majority of molecules in the gas phase.
b. The molecules with the strongest intermolecular attractions will have the lowest kinetic energy. Since molecules of A have the lowest kinetic energy, they will have the strongest intermolecular attractions.

c. The molecules with the strongest intermolecular attractions will have the lowest vapor pressure. Thus, molecules of A would have the lowest vapor pressure.

Conceptual Problem 11.22

Consider a substance X with a $\Delta H_{vap} = 20.3$ kJ/mol and $\Delta H_{fus} = 9.0$ kJ/mol. The melting point, freezing point, and heat capacities of both the solid and liquid X are identical to those of water.

a. If you placed one beaker containing 50 g of X at -10°C and another beaker with 50 g of H_2O at -10°C on a hot plate and started heating them, which material would reach the boiling point first?

b. Which of the materials from part a., X or H_2O, would completely boil away first?

c. On a piece of graph paper, draw the heating curve for H_2O and X. How do the heating curves reflect your answers from parts a. and b. of this problem?

Concept Target

• Integrate the information presented in heating curves, ΔH_{fus} and ΔH_{vap}, to explain experimental results.

Solution

You will need to compare the heats of fusion and vaporization of substance X ($\Delta H_{fus} = 9.0$ kJ/mol and $\Delta H_{vap} = 20.3$ kJ/mol) with the values for water, which are $\Delta H_{fus} = 6.01$ kJ/mol and $\Delta H_{vap} = 40.7$ kJ/mol. Comparing values shows that ΔH_{fus} is 1.5 times larger for substance X, and ΔH_{vap} is 2.0 times larger for H_2O.

Heating the substance, or water, from -10°C to the boiling point is a three step process. Step 1 is to heat the solid from -10°C to 0°C, the freezing point. The heat required for this step is equal to mass x specific heat capacity x temperature change. Step 2 is to melt the solid to liquid at 0°C. The heat required for this step is equal to moles x ΔH_{fus}. Step 3 is to heat the liquid from 0°C to 100°C. The heat required for this step is equal to mass x specific heat capacity x temperature change.

a. Since the masses, heat capacities, and temperature changes for water and for substance X are all equal, the heat required for step 1 and step 3 are the same for both. Since ΔH_{fus} is larger for substance X (per mole), step 2 will require more heat for substance X, and thus take longer. Therefore, H_2O will reach the boiling point first.

b. To completely boil away the substance, an additional step is required. Step 4 is to boil the liquid to vapor at 100°C. The heat required for this step is equal to moles x ΔH_{vap}. Since the ΔH_{vap} values are much larger than the ΔH_{fus} values, step 4 will require much more heat than step 2 for both substance X and H_2O. Since ΔH_{vap} is smaller for substance

X (per mole), step 4 will require less heat for substance X, and thus take less time. The total heat required for the four steps is directly proportional to the time it would take to completely boil away the substance. Steps 1 and 3 are the same for both. Step 2 takes 1.5 times as long for substance X, but step 4 takes 2.0 times as long for water. Since step 4 requires the most heat, water will require more time to complete this step, so substance X will boil away first.

c. The heating curves for substance X and for water are shown below.

Heating curve for water

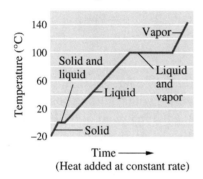

Heating curve for X

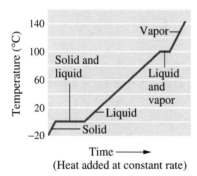

Conceptual Problem 11.23

Using the information presented in this chapter, explain why farmers spray water above and on their fruit trees on still nights when they know the temperature is going to drop below 0°C. (*Hint*: Totally frozen fruit is what the farmers are trying to avoid).

Concept Target

• Use a common application to illustrate that the phase change from liquid to solid, $-\Delta H_{fus}$, requires thermal energy.

Solution

The water that the farmers spray above and on their fruit will be warmer than the temperature of the fruit on the trees. Therefore, as the temperature of the air drops, it absorbs heat from the water, converting it into ice, before absorbing any heat from the fruit. The heat released when the liquid to solid phase change occurs prevents the fruit from freezing.

Conceptual Problem 11.24

You are presented with three bottles, each containing a different liquid: bottle A, bottle B, and bottle C. Bottle A's label states that it is an ionic compound with a boiling point of 35°C. Bottle B's label states that it is a molecular compound with a boiling point of 29.2°C. Bottle C's label states that it is a molecular compound with a boiling point of 67.1°C.
a. Which of the compounds is most likely to be incorrectly identified?
b. If Bottle A were a molecular compound, which of the compounds has the strongest intermolecular attractions?
c. If Bottle A were a molecular compound, which of the compounds would have the highest vapor pressure?

Concept Target

• Relate the physical properties of solids to their chemical composition and structure.

Solution

a. Bottle A is most likely mislabeled. If it is an ionic compound, the boiling point should be higher than 35°C. Most ionic compounds are solids, with high melting points.
b. The substance with the highest boiling point will have the strongest intermolecular attractions. Thus, the compound in bottle C has the strongest intermolecular attractions.
c. The substance with the lowest boiling point will have the highest vapor pressure. Thus the substance in bottle B will have the highest vapor pressure.

Conceptual Problem 11.25

Shown here is a representation of a unit cell for a crystal. The large balls are atom A (they are orange in text), and the small black balls are atom B.

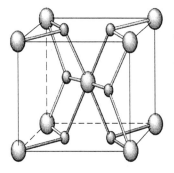

a. What is the chemical formula of the compound that has this unit cell ($A_?B_?$)?
b. Consider the configuration of the A atoms. Is this a cubic unit cell? If so, which type?

Concept Target

• Use the drawing of a unit cell to determine the chemical formula and the structure of a compound.

Solution

a. Considering the A atoms, there are nine per cell. Of these, eight are in corners and contribute 1/8 per cell, and one is completely inside the cell. Thus there are 8 x (1/8) + 1 x (1) = 2 A atoms per cell. Next, considering the B atoms, there are six per cell. Of these, four are in faces and contribute 1/2 per cell, and two are completely inside the cell. Thus, there are 4 x (1/2) + 2 x (1) = 4 B atoms per unit cell. The ratio of A atoms to B atoms is 2 to 4, or 1 to 2. Thus, the formula of the compound is AB_2.
b. The A atoms are in the arrangement of a body-centered cell.

Conceptual Problem 11.26

Assuming normal winter conditions (-1.5°C and 1.0 atm pressure), consult the phase diagram for water (Fig. 11.11) and come up with a reason why ice skates and sleds slide so well on solid water. Keep in mind that sleds and ice skates do not typically slide well on other solid surfaces (concrete and metal, for example).

Concept Target

• Extract information from a phase diagram to explain everyday observations.

Solution

Consulting the phase diagram for water (Figure 11.11), you see that by increasing the pressure on the solid ice at constant temperature, you will convert solid ice into liquid water. The film of liquid that forms allows the skates and sleds to slide so well.

Conceptual Problem 11.27

If you place room temperature water in a well-insulated cup and allow some of the water to evaporate, the temperature of the water in the cup will drop lower than room temperature. Come up with an explanation for this observation.

Concept Target

• Demonstrate that evaporation is a process that requires thermal energy and show that the surroundings are typically the source of this energy.

Solution

As the water evaporates, the molecules with higher kinetic energy escape the liquid, leaving behind the molecules with lower energy. The result is a drop in temperature of the liquid. Since the cup is well insulated, the energy lost with the evaporated molecules is not rapidly replaced.

Conceptual Problem 11.28

The heats of vaporization for water and carbon disulfide are 40.7 kJ/mol and 26.8 kJ/mol, respectively. A vapor (steam) burn occurs when the concentrated vapor of a substance condenses on your skin. Which of these substances, water or carbon disulfide, would result in the most severe burn if identical quantities of each vapor at a temperature just above their boiling point came in contact with your skin?

Concept Target

• Illustrate that the amount of energy that is released when a gas (vapor) condenses is a function of ΔH_{vap} which is directly related to the strength of the intermolecular attractions.

Solution

The heat released when the vapor of a substance condenses to liquid is equal to the negative (opposite) of the heat of vaporization for the substance. Due to its strong intermolecular attractions, water has a larger heat of vaporization, so it releases more heat when condensing on the skin causing a more severe burn.

Chapter 12

Solutions

Concept Check 12.1

Identify the solute(s) and solvent(s) in the following solutions.
a.   80 g of Cr and 5 g of Mo
b.   5 g of $MgCl_2$, dissolved in 1000 g of H_2O
c.   39% N_2, 41% Ar, and the rest O_2

Concept Target

• Identify the solute and solvent in a variety of solutions.

Solution

In each case, the component present in the greatest amount is the solvent.
a.   Mo is the solute, and Cr is the solvent.
b. $MgCl_2$ is the solute, and water is the solvent.
c. N_2 and O_2 are the solutes, and Ar is the solvent.

Concept Check 12.2

The hypothetical ionic compound, AB_2, is very soluble in water. Another hypothetical ionic compound, CB_2, is only slightly soluble in water. The lattice energies for these compounds are about the same. Provide an explanation for the solubility difference between these compounds.

Concept Target

• Explore how lattice energy and ionic size affect the solubility of ionic compounds in water.

Solution

The two main factors to consider when determining the solubility of an ionic compound in water are: ionic size and lattice energy. In this case the lattice energy for the two compounds is the same; you can discount its effects. Since a smaller cation will have a more concentrated electric field leading to a larger energy of hydration, you would expect AB_2 to have a greater energy of hydration: AB_2 is the more soluble compound.

Concept Check 12.3

Most fish have a very difficult time surviving at elevations much above 3500 m. How could Henry's law be used to account for this fact?

Concept Target

• Use the information presented in Henry's law to explain a "real world" observation.

Solution

As the altitude increases, the percent of oxygen in air decreases, and thus the partial pressure decreases. Above 3500 m, the partial pressure of oxygen in air has decreased to the point that not enough will dissolve in the water to sustain the fish.

Concept Check 12.4

You need to boil a water-based solution at a temperature lower than 100°C. What kind of liquid could you add to the water to make this happen?

Concept Target

• Highlight the relationship between the chemical composition of a solution and the vapor pressure of the solution.

Solution

In order to boil at a lower temperature than water, the vapor pressure of the solution (water + liquid) must be greater than water. In order to make this solution, you must add a liquid that is both soluble in water and chemically similar to water. It must have a higher vapor pressure than water and a boiling point lower than 100°C. One possible liquid is ethanol, with a boiling point of 78.3°C (Table 12.3).

Concept Check 12.5

Explain why pickles are stored in a brine (salt) solution. What would the pickles look like if they were stored in water?

Concept Target

• Recognize conditions where osmosis will occur and predict the outcome of an experiment that involves osmosis.

Solution

By the principle of osmosis, in brine solution, water will flow out of the pickle (lower concentration of ions) into the brine (higher concentration of ions). If the pickles were stored in a water solution, the water (lower concentration of ions) would flow into the pickle (higher concentration of ions) and cause it to swell up and probably burst.

Concept Check 12.6

Each of the following substances is dissolved in a separate 10.0-L container of water: 1.5 mol NaCl, 1.3 mol of Na_2SO_4, 2.0 mol $MgCl_2$, and 2.5 mol KBr. Without doing extensive calculations, rank the boiling points of each of the solutions from highest to lowest.

Concept Target

• Use solubility information and the colligative properties of ionic solutions (freezing point depression) to qualitatively predict the freezing points of various ionic solutions.

Solution

Each of these solutions is a water solution of identical volume (normal boiling point 100°C), containing a different number of moles of solute. The boiling point of a solution can be determined by the formula $\Delta T_b = iK_bm$. The solution with the largest ΔT_b will have the highest boiling point. Since K_b is a constant, this will be for the compound with the largest factor of $i \bullet m$. Also, since the volume is constant, the factor reduces to $i \bullet$ moles. Ideally, all of the compounds will dissolve completely, so NaCl and KBr have $i = 2$, and Na_2SO_4 and $MgCl_2$ have $i = 3$. This gives

> for NaCl, $i \bullet$ moles $= 2 \times 1.5 = 3.0$
> for Na_2SO_4, $i \bullet$ moles $= 3 \times 1.3 = 3.9$
> for $MgCl_2$, $i \bullet$ moles $= 3 \times 2.0 = 6.0$
> for KBr, $i \bullet$ moles $= 2 \times 2.0 = 4.0$

The result is given from highest boiling point to lowest boiling point:

$MgCl_2 > KBr > Na_2SO_4 > NaCl$

Concept Check 12.7

If electrodes that are connected to a direct current (DC) source are dipped into a beaker of colloidal iron(III) hydroxide, a precipitate collects at the negative electrode. Explain why this happens.

Concept Target

• Use the behavior and properties of colloidal particles to explain experimental observations.

Solution

Iron(III) hydroxide is a hydrophobic colloid. As the colloid forms in water, an excess of iron(III) ion (Fe^{3+}) is present on the surface, giving each crystal an excess of positive charge. These positively charged crystals repel one another, so aggregation to larger particles of iron(III) hydroxide is prevented. When the electrodes are dipped into the colloidal solution, iron(III) hydroxide precipitates because electrons from the negative electrode neutralize the excess positive charge on the iron(III) hydroxide, allowing larger particles to form (precipitate).

Conceptual Problem 12.21

Even though the oxygen demands of trout and bass are different, they can exist in the same body of water. However, if the temperature of the water in the summer gets above about 23°C, the trout begin to die, but not the bass. Why is this the case?

Concept Target

• Reinforce the relationship between the solubility of a gas and the temperature of a solution (Henry's law).

Solution

The amount of oxygen dissolved in water decreases as the temperature increases. Thus, at the lower temperatures, there is enough oxygen dissolved in the water to support both bass and trout. But, as the temperature rises above 23°C, there is not enough dissolved oxygen in the warm water to support the trout who need more O_2.

Conceptual Problem 12.22

You want to purchase a salt to melt snow and ice on your sidewalk. Which one of the following salts would best accomplish your task using the least amount: KCl, $CaCl_2$, PbS_2, $MgSO_4$, or AgCl?

Concept Target

• Use solubility information and the colligative properties of ionic solutions (freezing point depression) to qualitatively predict the freezing points of various ionic solutions.

Solution

The salt that would best accomplish the task would be the salt that lowers the freezing point of water the most. This in turn would be the salt with the largest i factor. Ideally, if each salt dissolved completely, KCl, $MgSO_4$, and AgCl would have $i = 2$. Similarly, $CaCl_2$ and PbS_2 would have $i = 3$. Of the latter two salts, $CaCl_2$ is more soluble than PbS_2, so its i factor is closer to 3. Therefore, the salt with the largest i factor is $CaCl_2$, so it would lower the freezing point of water the most and would best accomplish the task.

Conceptual Problem 12.23

Ten grams of the hypothetical ionic compounds XZ and YZ are each placed in a separate 2.0 L beaker of water. XZ completely dissolves, whereas YZ is insoluble. The energy of hydration of the Y^+ ion is greater than the X^+ ion. Explain this difference in solubility.

Concept Target

• Explore how lattice energy and ionic size affect the solubility of ionic compounds in water.

Solution

The two main factors to consider when determining the solubility of an ionic compound in water are ionic size and lattice energy. Ionic size is inversely related to the energy of hydration; the smaller the ion, the greater the energy of hydration. Keep in mind that the greater the energy of hydration, the more likely it is for a compound to dissolve. The amount of lattice energy is directly related to the solubility of the compound; the lower the lattice energy, the more likely it is for the compound to dissolve.

Taking into account these factors, in order to increase the solubility of a compound you need to decrease the ionic size and decrease the lattice energy. Since the energy of hydration of the Y+ ion is greater than that of the X+ ion (making XZ less soluble), in order for XZ to be more soluble than YZ the lattice energy must be less for the XZ compound.

Conceptual Problem 12.24

Small amounts of a nonvolatile, nonelectrolyte solute and a volatile solute are each dissolved in separate beakers containing 1 kg of water. If the number of moles of each solute is equal:
a.   Which solution will have the higher vapor pressure?
b.   Which solution will boil at a higher temperature?

Concept Target

• Understand how the volatility of a solute affects the vapor pressure of a solution.

Solution

a.   According to Raoult's law, the addition of a nonvolatile, nonelectrolyte to a solvent will lower the vapor pressure of the solvent, so we would expect the vapor pressure of such a solution to be lower than that of the pure solvent (water in this case). When a volatile solute is added to a solvent, the vapor pressure of the solution is dependent upon the mole fraction of the solute and solvent and the vapor pressures of both the solute and solvent. Since the solute is volatile (a high vapor pressure relative to water), the solution must have a higher vapor pressure than pure water.
b.   Keeping in mind that a solution will boil when the vapor pressure equals the pressure pushing on the surface of the solution, the solution with the greater vapor pressure will boil at a lower temperature. In this case, it is the solution with the volatile solute.

Conceptual Problem 12.25

A Cottrell precipitator consists of a column containing electrodes that are connected to a high-voltage direct current (DC) source. The Cottrell precipitator is placed in smokestacks to remove smoke particles from the gas discharged from an industrial plant. Explain how you think this works.

Concept Target

• Use the behavior and properties of colloidal particles to explain how a common industrial process works.

Solution

Smoke particles carry a small net charge, preventing them from forming larger particles that would settle to the bottom of the smokestack. The charged smoke particles are neutralized by the current which then allows them to aggregate into large particles. These large particles are too big to be carried out of the stack.

Conceptual Problem 12.26

Consider the following dilute NaCl(aq) solutions.

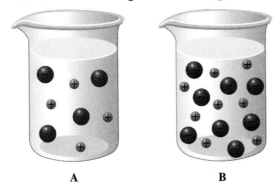

A   B

a. Which one will boil at a higher temperature?
b. Which one will freeze at a lower temperature?
c. If the solutions were separated by a semipermeable membrane that allowed only water to pass, which solution would you expect to show an increase in the concentration of NaCl?

Concept Target

• Use molecular models to develop a qualitative understanding of the colligative properties.

Solution

a. Since beaker B contains more solute particles, according to Raoult's law, it will boil at a higher temperature than beaker A.
b. More particles in solution lead to a lower vapor pressure, which in turn lowers the freezing point of a solution. Since beaker B contains more solute particles, it will freeze at a lower temperature than beaker A.
c. When separated by a semipermeable membrane, the solvent from the less concentrated solution flows into the more concentrated solution. Because of this, the water will flow from beaker A to beaker B causing an increase in the concentration of NaCl in beaker A.

Conceptual Problem 12.27

A green leafy salad wilts if left too long in a salad dressing containing vinegar and salt. Explain what happens.

Concept Target

• Use osmosis to explain a commonly observed phenomenon.

Solution

Vinegar is a solution of acetic acid (solute) and water (solvent). Because the salt concentration outside the lettuce leaf is higher than inside, water will pass out of the lettuce leaf into the dressing via osmosis. The result is that the lettuce will become wilted.

Conceptual Problem 12.28

People have proposed towing icebergs to arid parts of the earth as a way to deliver fresh water. Explain why icebergs do not contain salts although they are formed by the freezing of ocean water (i.e., saltwater).

Concept Target

• Realize that the pure solute freezes out of a solution first.

Solution

As a solution freezes, pure solvent forms without any of the solute present. This means that as ocean water freezes to make icebergs, it freezes as pure water without the salt present.

Chapter 13

Materials of Technology

Concept Check 13.1

Why must a metal containing mineral be reduced to obtain the free metal?

Concept Target

• Emphasize that metals are usually found in nature in an oxidized state.

Solution

Metals in ores and minerals are in an oxidized form where the free metal form is the neutral, elemental state.

Concept Check 13.2

A semi-conductor is a material where there is a separation between the filled band and an unfilled band called an energy gap. Would it take more or less energy to make a semi-conductor carry current?

Concept Target

• Apply band theory concepts to predict the electrical behavior and properties of a semi-conductor.

Solution

It would take more energy because you would need to give the electrons enough energy to jump the gap into the unoccupied orbitals that make up the unfilled band.

Concept Check 13.3

For certain chemical reactions, quartz containers are used instead of glass. What would the meniscus of water look like in a quartz test tube? Come up with an explanation for your answer.

Concept Target

- Integrate chemical information from this and previous chapters to predict the outcome of an experiment.

Solution

The meniscus of water would look the same as in glass, curved downward from the walls of the container. It attains this shape because of the hydrogen bonding interactions between the O atoms in the SiO_2 tetrahedra and the H atoms of the water molecules.

Conceptual Problem 13.29

Unless zinc is purified, cadmium is normally an impurity in the metal. Why might you expect this to be the case.

Concept Target

- Investigate ore refining.

Solution

The refinement process for zinc produces metal that contains impurities such as lead, cadmium, and iron. Zinc and cadmium share many similar chemical and physical properties. They are both in group IIB on the periodic table.

Conceptual Problem 13.30

Electrolysis is used to obtain some metals from their compounds. List some metals obtained this way.

Concept Target

- Integrate information on the production of metals presented throughout the chapter.

Solution

Metals obtained by electrolysis from their compounds include lithium, sodium, magnesium, and aluminum.

Conceptual Problem 13.31

Aluminum is the third most abundant element (first most abundant metal) in the earth's crust. Does this mean that aluminum ores are widespread and plentiful? Explain.

Concept Target

- Illustrate the fact that although many elements are abundant on earth, this does not ensure that large quantities are readily available for human use.

Solution

No, most aluminum is found in aluminum-containing clays and not in a mine. Also, much bauxite occurs in tropical and subtropical regions, where mining is not easy.

Conceptual Problem 13.32

The text says that the higher density of diamond compared with graphite suggests that the application of higher pressure would facilitate the transformation of graphite to diamond. Explain the reasoning behind this statement.

Concept Target

- Make connections between the structure of graphite and diamonds and their physical properties.

Solution

The higher density of diamond structure graphite indicates that the diamond structure of carbon contains more atoms per unit volume than the graphite structure. Therefore, increasing the pressure on graphite would increase the number of carbon atoms per unit volume (the carbon atoms are now packed closer together in a more dense structure). This increase in density (reduction of the distance between carbon atoms) then helps each carbon atom form four covalent bonds to neighboring carbon atoms, thereby forming a diamond.

Conceptual Problem 13.33

Diamond is an insulator, but when small amounts of boron are added it becomes a conductor. What is the explanation for this change in conduction? What are the electric current carriers?

Concept Target

• Understand the effect of doping on insulators.
• Make the distinction between p-type and n-type semiconductors.

Solution

Boron has one less electron than carbon, so it acts like a p-type semiconductor in that positive holes are created. Since positive holes are created, the electric current carriers are the positive holes.

Conceptual Problem 13.34

Is a quartz crystal (SiO_2) a mineral or a rock?

Concept Target

• Highlight the definition of a mineral.

Solution

Because it has definite chemical composition, quartz crystal (SiO_2) is a mineral.

Conceptual Problem 13.35

Orthoclase feldspar has the chemical formula $KAlSi_3O_8$. Think about what other cations could replace those in this chemical formula and write the resulting formula for two such minerals.

Concept Target

• Use the periodic table to find chemically similar elements.

Solution

Examples are $MgCaSi_3O_8$ and $KNaMgSi_3O_8$.

Conceptual Problem 13.36

Cutting wheels have been made from alumina containing fine fibers of silicon carbide. How do the silicon carbide fibers help alumina in this application?

Concept Target

• Investigate the properties of ceramics.

Solution

The fibers of silicon carbide function as a hard, durable and abrasive material.

Chapter 14

Rates of Reaction

Concept Check 14.1

Shown here is a plot of the concentration of a reactant D versus time.

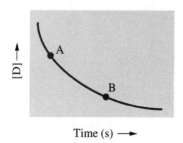

Time (s) →

a. How do the instantaneous rates at points A and B compare?

b. Is the rate for this reaction constant at all points in time?

Concept Target

• Use a graph of kinetics data to predict relative reaction rates.

Solution

a. Since the slope is steeper at point A, point A must be a faster instantaneous rate.

b. Since the curve is not a flat line, the rate of the reaction must be constantly changing over time. Therefore, the rate for the reaction cannot be constant at all points in time.

Concept Check 14.2

Consider the reaction $Q + R \rightarrow S + T$ and the rate law for the reaction: Rate = $k [Q]^0[R]^2$.

a. You run the reaction three times, each time starting with $[R] = 2.0\ M$. For each run you change the starting concentration of $[Q]$: run 1, $[Q] = 0.0\ M$; run 2, $[Q] = 1.0\ M$; run 3, $[Q] = 2.0\ M$. Rank the rates of the three reactions using each of these concentrations.

b. The way the rate law is written in this problem is not typical for expressions containing reactants that are zero order in the rate law. Write the rate law in the more typical fashion.

Concept Target

• Use a rate law and experimental data to qualitatively predict the outcome of a reaction.

Solution

a. Keeping in mind that all reactant species must be present in some concentration for a reaction to occur, the reaction with $[Q] = 0$ is the slowest since no reaction occurs. The other two reactions are equal in rate because the reaction is zero order with respect to $[Q]$: as long as there is some amount of Q present, the reaction rate depends on the $[R]$ which is constant in this case.

b. Since $[Q]^0 = 1$, you can rewrite the rate law as follows: Rate $= k[R]^2$.

Concept Check 14.3

Rate laws are not restricted to chemical systems; they are used to help describe many "everyday" events. For example, a rate law for tree growth might look something like this:

$$\text{Rate of Growth} = (\text{soil type})^w(\text{temperature})^x(\text{light})^y(\text{fertilizer})^z.$$

In this equation, like chemical rate equations, the exponents need to be determined by experiment. (Can you think of some other factors?)

a. Say you are a famous physician trying to determine the factors that influence the rate of aging in humans. Develop a rate law, like the one above, that would take into account at least four factors that affect the rate of aging.

b. Explain what you would need to do in order to determine the exponents in your rate law.

c. Consider smoking to be one of the factors in your rate law. You conduct an experiment and find that a person smoking two packs of cigarettes a day quadruples (4X) the rate of aging over that of a one-pack-a-day smoker. Assuming that you could hold all other factors in your rate law constant, what will be the exponent of the smoking term in your rate law?

Concept Target

• Emphasize that rate laws are not unique to chemical systems and that the concepts associated with rate laws have wide application.

Solution

a. A possible rate law is: Rate of Aging $= (\text{diet})^w(\text{exercise})^x(\text{sex})^y(\text{occupation})^z$. Your rate law probably will be different; however, the general form should be the same.

b. You would need a sample of people that have all of the factors the same except one. For example, using the equation given in part a., you could determine the effect of diet if you had a sample of people that were the same sex, exercised the same amount, and had the same occupation. You would need to isolate each factor in this fashion to determine the exponent on each factor.

c. The exponent on the smoking factor would be 2 since you see a fourfold rate increase: $[2]^2 = 4$.

Concept Check 14.4

A reaction believed to be either first or second order has a half-life of 20 s at the beginning of the reaction but a half-life of 40 s some time later. What is the order of the reaction?

Concept Target

• Illustrate that the half-life of a first order reaction is constant and varies with time when the reaction is second order.

Solution

The half-life of a first-order reaction is constant over the course of the reaction. The half-life of a second-order reaction depends on the initial concentration and becomes larger as time elapses. Thus, the reaction must be second order because the half-life increases from 20 s to 40 s after time has elapsed.

Concept Check 14.5

Considering the following potential energy curves for two different reactions:

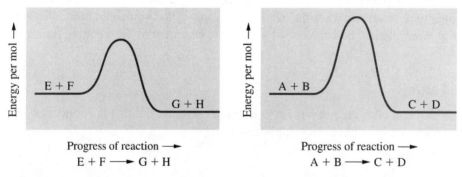

a. Which reaction has a higher activation energy for the forward reaction?
b. If both reactions were run at the same temperature and have the same orientation requirements to react, which one would have the larger rate constant?
c. Are these reactions exothermic or endothermic?

Concept Target

• Illustrate that reaction coordinate diagrams provide information on the activation energy and hence the rate constant for chemical reactions.

Solution

a. Since the "hump" is larger, the A + B reaction has a higher activation energy.
b. Since the activation energy is lower, the E + F reaction would have the larger rate constant. Keep in mind the inverse relationship between the activation energy, E_a, and the rate constant, k.
c. Since in both cases energy per mole of the reactants is greater than the products, both reactions are exothermic.

Concept Check 14.6

You are a chemist in charge of a research laboratory that is trying to increase the reaction rate for the balanced chemical reaction: $X + 2Y \rightarrow Z$.

a. One of your researchers comes into your office and states that she has found a material that significantly lowers the activation energy of the reaction. Explain the effect this will have on the rate of the reaction.
b. Another researcher states that after doing some experiments, he has determined that the rate law is rate = $k[X][Y]$. Is this possible?
c. Yet another person in the lab reports that the mechanism for the reaction is:

$2Y \rightarrow I$ (slow)
$X + I \rightarrow Z$ (fast)

Is the rate law from part b. consistent with this mechanism? If not, what should the rate law be?

Concept Target

• Highlight that the rate law of a chemical reaction must be consistent with the proposed mechanism for the reaction.

Solution

a. Her finding should increase the rate since the activation energy, E_a, is inversely related to the rate constant, k; a decrease in E_a results in an increase in the value of k.
b. This is possible because the rate law does not have to reflect the overall stoichiometry of the reaction.
c. No. Since the rate law is based on the slow step of the mechanism, it should be Rate = $k[Y]^2$.

Conceptual Problem 14.23

Consider the reaction: $3A \rightarrow 2B + C$.

a. One rate expression for the reaction is: Rate of formation of $C = + \dfrac{\Delta[C]}{\Delta t}$.

   Write two other rate expressions for this reaction in this form.

b. Using your two rate expressions, if you calculated the average rate of the reaction over the same time interval, would the rates be equal?

c. If your answer to part b. was no, write two rate expressions that would give an equal rate when calculated over the same time interval.

Concept Target

• Highlight the relationship between the stoichiometry of a reaction and the reporting of its rate.

Solution

a. You can write the rate expression in terms of the depletion of A:

   Rate of depletion of $A = - \dfrac{\Delta[A]}{\Delta t}$.

   Or, you can write the rate expression in terms of the formation of B:

   Rate of formation of $B = + \dfrac{\Delta[B]}{\Delta t}$.

b. No. Consider the stoichiometry of the reaction which indicates that the rate of depletion of A would be faster than the rate of formation of B: for every three moles of A that are consumed, two moles of B would be formed.

c. Taking into account the stoichiometry of the reaction, the two rate expressions that would give an equal rate when calculated over the same time interval are

   $$Rate = - \dfrac{\Delta[A]}{3\Delta t} = \dfrac{\Delta[B]}{2\Delta t}.$$

Conceptual Problem 14.24

Given the reaction $2A + B \rightarrow C + 3D$, can you write the rate law for this reaction? If so, write the rate law; if not, why?

Concept Target

• Focus on the fact that a rate law only can be determined by experiment, not by the stoichiometry of the reaction.

Solution

You cannot write the rate law for this reaction from the information given. The rate law can only be determined by experiment, not by the stoichiometry of the reaction.

Conceptual Problem 14.25

You perform some experiments for the reaction $A \rightarrow B + C$ and determine the rate law has the form: Rate $= k[A]^x$. Calculate the value of the exponent x for each of the following cases.
a. [A] is tripled and you observe no rate change.
b. [A] is doubled and the rate doubles.
c. [A] is tripled and the rate goes up by a factor of 27.

Concept Target

• Explore the relationship between the value of the exponent in a rate law and experimental observations.

Solution

a. If the concentration is tripled but there is no effect on the rate, the order of the reaction must be zero. Thus, $x = 0$.
b. If the concentration is doubled and the rate doubles, it is a first order reaction. Thus, $x = 1$.
c. If the concentration is tripled and the rate goes up by a factor of 27, it is a third order reaction. Thus, $x = 3$.

Conceptual Problem 14.26

A friend of yours runs a reaction and generates the following plot. She explains that in following the reaction, she measured the concentration of a compound that she calls "E."

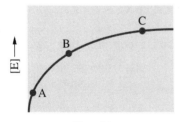

Time (s) ⟶

a. Your friend tells you that E is either a reactant or product. Which is it and why?
b. Is the average rate faster between points A and B or B and C? Why?

Concept Target

• Underscore that rate information can be extracted from an experimental concentration versus time plot.

Solution

a. E must be a product since its concentration increases with time. If E were a reactant, you would expect the concentration to decrease over time.
b. The average rate is faster between points A and B since the slope of the curve is steeper in this region. Remember, the steeper the curve, the greater the rate of change.

Conceptual Problem 14.27

Given the hypothetical plot shown here for the concentration of compound Y versus time, answer the following questions.

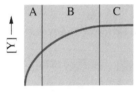

Time (s) ⟶

a. In which region of the curve does the rate have a constant value (A, B, or C)?
b. In which region of the curve is the rate the fastest (A, B, or C)?

Concept Target

• Illustrate that rate information can be extracted from an experimental concentration versus time plot.

Solution

a. The rate has a constant value in region C, since the slope of the curve is constant (flat) in this region.
b. The rate is the fastest in region A, since the slope of the curve is steepest in this region.

Conceptual Problem 14.28

You carry out the following reaction by introducing N_2O_4 into an evacuated flask and observing the concentration change of the product over time.

$$N_2O_4(g) \rightarrow 2NO_2(g)$$

Which one of the curves shown here reflects the data collected for this reaction?

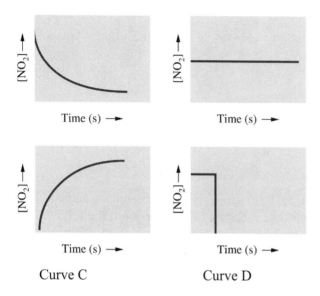

Curve C   Curve D

Concept Target

• Predict the shape of a concentration versus time plot for the products of a chemical reaction.

Solution

Since NO_2 is a product in the reaction, its concentration must increase with time. The only graph that has $[NO_2]$ increasing with time is curve C.

Conceptual Problem 14.29

You are running the reaction $2A + B \rightarrow C + 3D$. Your lab partner has conducted the first two experiments to determine the rate law for the reaction. He has recorded the initial rates for these experiments in another data table. Come up with some reactant concentrations for Experiment 3 that will allow you to determine the rate law by measuring the initial rate.

Experiment Number	Concentration of A (M)	Concentration of B (M)
1	1.0	1.0
2	2.0	1.0
3		

Concept Target

• Understand the experimental conditions that are necessary to collect meaningful kinetics data.

Solution

A number of answers will work as long as you match one of the existing concentrations of A or B. For example: [A] = 2.0 M with [B] = 2.0 M, or [A] = 1.0 M with [B] = 2.0 M.

Conceptual Problem 14.30

The chemical reaction A → B + C has a rate constant that obeys the Arrhenius equation. Predict what happens to both the rate constant k and the rate of the reaction if the following were to occur:

a.   a decrease in temperature.
b.   an increase in the activation energy of the forward and reverse reactions.
c.   an increase in both activation energy and temperature.

Concept Target

• Use the relationships established in the Arrhenius equation to qualitatively predict how temperature changes and activation energy affect the rates of chemical reactions.

Solution

The Arrhenius equation is $k = Ae^{-E_a/RT}$

a.   When the temperature is decreased, the rate constant, k, will also decrease. When k decreases, the rate also decreases.
b.   When the activation energy is increased, the rate constant, k, also decreases. When k decreases, the rate also decreases.
c.   Since the activation energy is in the numerator and the temperature is in the denominator, you cannot predict the effect without knowing the magnitude of the changes.

Chapter 15

Chemical Equilibrium

Concept Check 15.1

Two substances A and B react to produce substance C. When reactant A decreases by an amount x, product C increases by amount x. When reactant B decreases by an amount x, product C increases by amount $2x$. Write the chemical equation for the reaction.

Concept Target

• Underscore the relationship between the coefficients in a chemical equation and the amounts of substances involved in a reaction.

Solution

The statement that when reactant A decreases by an amount x, product C increases by amount x implies that A and C have the same coefficients. The statement that when reactant B decreases by an amount x, product C increases by amount $2x$ implies that the coefficient of C is twice that of B. Therefore, the coefficient of A is twice that of B. The simplest equation satisfying these conditions is $2A + B \rightarrow 2C$.

Concept Check 15.2

Carbon monoxide and hydrogen react in the presence of a catalyst to form methanol, CH_3OH:

$$CO(g) + 2H_2(g) \rightleftharpoons CH_3OH(g).$$

An equilibrium mixture of these three substances is suddenly compressed so that the concentrations of all substances initially double. In what direction does the reaction go as a new equilibrium is attained?

Concept Target

• Understand better how to use the reaction quotient to decide the direction of a reaction.

Solution

The concentration of each substance initially doubles. This means that each concentration factor in the reaction quotient expression is double that in the initial equilibrium mixture. Because this expression contains $[CO][H_2]^2$ in the denominator, the denominator increases by a factor of 2^3. However, the numerator contains only $[CH_3OH]$, which merely doubles. So, the reaction quotient equals 2 divided by 2^3, or ¼, times the equilibrium constant. To approach the equilibrium constant, the numerator of the reaction quotient must increase and the denominator must decrease. This means that more CH_3OH must be produced. The reaction goes from left to right.

Concept Check 15.3

A and B react to produce C according to the chemical equation: $A + B \rightarrow C$. Enough A and B are added to an equilibrium reaction mixture of A, B, and C so that when equilibrium is again attained the amounts of A and B are doubled in the same volume. How is the amount of C changed?

Concept Target

• Apply the concept of the equilibrium constant in a nonnumerical example.

Solution

The equilibrium-constant expression is $[C]/([A][B])$. A new equilibrium is attained in which the equilibrium-constant expression is $[C]'/([A]'[B]') = [C]'/(2[A]2[B])$, where primes indicate new equilibrium concentrations. The value of the equilibrium-constant expression, though, must remain fixed in value, so $[C]'/(2[A]2[B])$ equals $[C]/([A][B])$. This means that the new concentration of C, or $[C]'$, must be 4 times larger than the original equilibrium value. Thus, the concentration of C is quadrupled.

Conceptual Problem 15.13

During an experiment with the Haber process, a researcher put 1 mol N_2 and 1 mol H_2 into a reaction vessel to observe the equilibrium formation of ammonia, NH_3.

$$N_2(g) + 3H_2(g) \rightleftharpoons 2NH_3(g)$$

When these reactants come to equilibrium, assume that x mol H_2 react. How many moles of ammonia form?

Concept Target

• Underscore the relationship between the coefficients in a chemical equation and the amounts of substances involved in a reaction.

Solution

For each three moles of H_2 that react, two moles of ammonia form. The mole ratio is

$$\frac{2 \text{ mol NH}_3}{3 \text{ mol H}_2}$$

If x mol H_2 react, the amount of ammonia that forms is

$$x \text{ mol H}_2 \times \frac{2 \text{ mol NH}_3}{3 \text{ mol H}_2} = \frac{2x}{3} \text{ mol NH}_3.$$

Conceptual Problem 15.14

Suppose liquid water and water vapor exist in equilibrium in a closed container. If you add a small amount of liquid water to the container, how does this affect the amount of water vapor in the container? If, instead, you add a small amount of water vapor to the container, how does this affect the amount of liquid water in the container?

Concept Target

• Examine heterogeneous equilibrium in a qualitative context.

Solution

The addition of a pure liquid does not affect an equilibrium (the pure liquid does not appear in the equilibrium constant; in effect, the concentration of the liquid does not change). Thus, the amount of water vapor does not change appreciably (more precisely, the amount of vapor decreases slightly because the liquid takes up more room in the container). If, instead, you add water vapor to the container, vapor condenses until the original vapor pressure is restored. Thus, the amount of liquid water in the container increases.

Conceptual Problem 15.15

A mixture initially consisting of 2 mol CO and 2 mol H_2 comes to equilibrium with methanol, CH_3OH, as the product:

$$CO(g) + 2H_2(g) \rightleftharpoons CH_3OH(g)$$

At equilibrium, the mixture will contain which of the following?
a. less than 1 mol CH_3OH
b. 1 mol CH_3OH
c. more than 1 mol CH_3OH but less than 2 mol
d. 2 mol CH_3OH
e. more than 2 mol CH_3OH

Concept Target

• Explore the effect of a limiting reactant on the amount of product in a chemical equilibrium.

Solution

Hydrogen, H_2, is the limiting reactant, so the maximum amount of CH_3OH that could form is 1 mol. However, because the reaction comes to equilibrium before it can go to completion, less than 1 mol of CH_3OH forms. The answer is a.

Conceptual Problem 15.16

When a continuous stream of hydrogen gas, H_2, passes over hot magnetic iron oxide, Fe_3O_4, metallic iron and water vapor form. When a continuous stream of water vapor passes over hot metallic iron, the oxide Fe_3O_4 and H_2 form. Explain why the reaction goes in one direction in one case but in the reverse direction in the other.

Concept Target

• Understand that an excess of any one substance in an equilibrium will push a reaction in the opposite direction.

Solution

The system must exist as an equilibrium mixture of all four substances. The reaction can be represented as

$$Fe_3O_4(s) + 4H_2(g) \rightleftharpoons 3Fe(s) + 4H_2O(g)$$

If you pass $H_2(g)$ over iron oxide, the reaction shifts to the right and metallic iron and water vapor form. If, instead, you pass water vapor over metallic iron, the reaction shifts to the left, and iron oxide and H_2 will form. An excess of one reactant pushes the reaction in the opposite direction.

Conceptual Problem 15.17

During the commercial preparation of sulfuric acid, sulfur dioxide reacts with oxygen in an exothermic reaction to produce sulfur trioxide. In this step, sulfur dioxide mixed with oxygen-enriched air passes into a reaction tower at about 420°C, where reaction occurs on a vanadium(V) oxide catalyst. Discuss the conditions used in this reaction in terms of its effect on the yield of sulfur trioxide. Are there any other conditions that you might explore in order to increase the yield of sulfur trioxide?

Concept Target

• Explore the effect of reaction conditions on the yield of a product.

Solution

The formation of SO_3 can be represented by the following reaction:

$$2SO_2(g) + O_2(g) \rightleftharpoons 2SO_3(g)$$

The first condition mentioned is the oxygen-enriched air. The higher concentration of O_2 drives the reaction to the right, increasing the yield of SO_3. Next, the reaction temperature is 420°C. Since the reaction is exothermic, the elevated temperature drives the reaction to the left, decreasing the yield of SO_3. The vanadium(V) oxide catalyst affects the rate of the reaction, but not the amount of SO_3 that forms at equilibrium. Finally, the SO_3 that forms in the reaction is absorbed by concentrated sulfuric acid and removed from the system. This causes the reaction to shift to the right, forming more SO_3.

Another condition that can be explored is a change in pressure. Since there are more moles of gas on the left side, an increase in pressure (or decrease in volume) should cause more SO_3 to form.

Conceptual Problem 15.18

The following reaction is carried out at 500 K in a container equipped with a movable piston.

$$A(g) + B(g) \rightleftharpoons C(g); K_c = 10 \text{ (at 500 K)}$$

After the reaction has reached equilibrium, the container has the composition depicted here. (In order to arrive at the correct answer(s), viewing the color text version of the figure associated with this problem is advisable.)

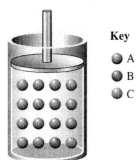

Key

● A
● B
● C

Suppose the container volume is doubled.
a. How does the equilibrium composition shift?
b. How does the concentration of each of the reactants and the product change? (That is, does the concentration increase, decrease, or stay the same?)

Concept Target

• Determine how pressure changes affect a reaction at equilibrium.

Solution

a. When the volume is doubled, the pressure is then reduced by one-half. A decrease in pressure results in the equilibrium shifting to the side of the reaction with the greater number of moles of gas. Note that in this reaction, there are 2 moles of gas reactants versus 1 mole of gas products. Since the pressure has been reduced, the reaction will shift to the left, towards the side of the reaction with the greater number of moles of gas.
b. Since the reaction has been shifted to the left, the concentrations of A and B will increase, while the concentration of C will decrease.

Conceptual Problem 15.19

An experimenter introduces 4.0 mol of gas A into a 1.0-L container at 200 K to form product B according to the reaction

$$2A(g) \rightleftharpoons B(g)$$

Using the following data collected during the experiment, calculate the equilibrium constant at 200 K.

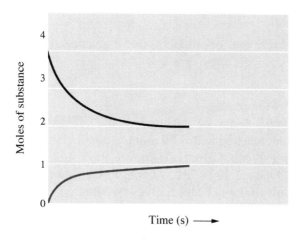

Concept Target

- Interpret a graph of experimental data to determine the equilibrium constant for a reaction.

Solution

Equilibrium has been reached when the concentration of reactants and products is constant. The equilibrium region on the graph is where the lines flatten out indicating that the concentrations of reactants and products are not changing. At equilibrium, the concentration of A is approaching 2.0 M, and the concentration of B is approaching 1.0 M. Substituting into the equilibrium constant expression gives

$$K_c = \frac{[B]}{[A]^2} = \frac{1.0}{(2.0)^2} = 0.25$$

Conceptual Problem 15.20

At some temperature, a 100-L reaction vessel contains a mixture that is initially 1.00 mol CO and 2.00 mol H_2. The vessel also contains a catalyst so that the following equilibrium is attained:

$$CO(g) + 2H_2(g) \rightleftharpoons CH_3OH(g)$$

At equilibrium, the mixture contains 0.100 mol CH_3OH. In a later experiment in the same vessel, you start with 1.00 mol CH_3OH. How much methanol is there at equilibrium? Explain.

Concept Target

• Understand that equivalent amounts of either reactants or products yield the same equilibrium mixture.

Solution

If you start with 1.00 mol CH_3OH, there will be 0.100 mol CH_3OH present at equilibrium. Assume that the reaction could go to completion. Starting with 1.00 mol CO and 2.00 mol H_2, stoichiometry shows that you would obtain 1.00 mol CH_3OH. Therefore, 1.00 mol CH_3OH is chemically equivalent to 1.00 mol CO and 2.00 mol H_2. At equilibrium, starting either with 1.00 mol CO and 2.00 mol H_2 or with 1.00 mol CH_3OH gives the same equilibrium mixture, one containing 0.100 mol CH_3OH.

Chapter 16

Acids and Bases

Concept Check 16.1

Chemists in the seventeenth century discovered that the substance that gives red ants their irritating bite is an acid with the formula $HCHO_2$. They called this substance formic acid after the ant, whose Latin name is *Formica rufus*. Formic acid has the following structural formula and molecular model.

$$\underset{\displaystyle H-\overset{\displaystyle O}{\overset{\|}{C}}-O-H}{}$$

Write the acid-base equilibria connecting all components in the aqueous solution. Now list all of the species present.

Concept Target

• Understand the equilibria that occur in a solution of weak acid and the species that result.

Solution

In any aqueous solution, you should consider the autoionization of water. And because we have a solution of a weak acid in water, you should also consider the equilibrium between this acid and water. Here are the two equilibria:

$H_2O(l) + H_2O(l) \rightleftharpoons H_3O^+(aq) + OH^-(aq)$

$HCHO_2(aq) + H_2O(l) \rightleftharpoons CHO_2^-(aq) + H_3O^+(aq)$

The species present in these equilibria are: $H_2O(l)$, $H_3O^+(aq)$, $OH^-(aq)$, $HCHO_2(aq)$, and $CHO_2^-(aq)$.

Concept Check 16.2

Formic acid, $HCHO_2$, is a stronger acid than acetic acid, $HC_2H_3O_2$. Which is the stronger base, formate ion, CHO_2^-, or acetate ion, $C_2H_3O_2^-$?

Concept Target

• Emphasize the relationships between acid strengths and conjugate base strengths.

Solution

The stronger acid gives up its proton more readily, and therefore its conjugate base ion holds onto a proton less strongly. In other words, the stronger acid has the weaker conjugate base. Because formic acid is the stronger acid, the formate ion is the weaker base. Acetate ion is the stronger base.

Concept Check 16.3

You have solutions of NH_3, HCl, NaOH, and $HC_2H_3O_2$ (acetic acid), all with the same solute concentrations. Rank these solutions in order of pH, from the highest to lowest.

Concept Target

• Obtain an understanding of the relationships between acid and base strength and pH.

Solution

In order to qualitatively answer this problem, it is essential that all of the solutions have the same solute concentrations. Bases produce solutions of pH greater than 7 where acids produce solutions of pH less than 7. NH_3 and NaOH are bases, and HCl and $HC_2H_3O_2$ are acids. NaOH is a stronger base than NH_3, so the NaOH solution would have the highest pH followed by the NH_3 solution. $HC_2H_3O_2$ is a much weaker acid than HCl so the $HC_2H_3O_2$ solution would have a higher pH than the HCl solution. Therefore, the ranking from highest to lowest pH for solutions with the same solute concentrations is: NaOH > NH_3 > $HC_2H_3O_2$ > HCl.

Conceptual Problem 16.15

Aqueous solutions of ammonia, NH_3 were once thought to be solutions of an ionic compound ammonium hydroxide, NH_4OH, in order to explain how solutions could contain hydroxide ion. Using the Brønsted-Lowry concept, show how NH_3 yields hydroxide ion in aqueous solution without involving the species NH_4OH.

Concept Target

• Underscore the role of Brønsted theory in understanding bases.

Solution

It is not necessary to have the species NH_4OH in order to have OH^- in the solution. When ammonia reacts with water, hydroxide ion forms in the reaction.

$$NH_3(aq) + H_2O(l) \rightleftharpoons NH_4^+(aq) + OH^-(aq)$$

Conceptual Problem 16.16

Blood contains several substances that minimize changes in its acidity by reacting with either an acid or a base. One of these is the hydrogen phosphate ion, HPO_4^{2-}. Write one equation showing this species acting as a Brønsted–Lowry acid and another in which the species acts as a Brønsted–Lowry base.

Concept Target

• Realize that some species can act as either acids or bases.

Solution

A reaction where HPO_4^{2-} acts as an acid is

$$HPO_4^{2-}(aq) + OH^-(aq) \rightleftharpoons PO_4^{3-}(aq) + H_2O(l).$$

A reaction where HPO_4^{2-} acts as a base is

$$HPO_4^{2-}(aq) + H_3O^+(aq) \rightleftharpoons H_2PO_4^-(aq) + H_2O(l).$$

Conceptual Problem 16.17

Self-contained environments, such as that of a space station, require that the carbon dioxide exhaled by people be continuously removed. This can be done by passing the air over solid alkali hydroxide, in which carbon dioxide reacts with hydroxide ion. What ion is produced by the addition of OH^- ion to CO_2? Use the Lewis concept to explain this.

Concept Target

• Strengthen understanding of the Lewis concept of acids and bases.

Solution

The hydroxide ion acts as a base and donates a pair of electrons to the O atom, forming a bond with CO_2 to give HCO_3^-.

Conceptual Problem 16.18

Compare the structures of HNO_2 and H_2CO_3. Which would you expect to be the stronger acid? Explain your choice.

Concept Target

• Note the relationship between acid strength and electronegativity of the central atom.

Solution

Nitrogen has greater electronegativity than carbon. You would expect the H—O bond in the H—O—N group to be more polar (with the H atom having a positive partial charge) than the H—O bond in the H—O—C group. Thus, based on their structure, you would expect HNO_2 to be the stronger acid.

Conceptual Problem 16.19

The value of the ion-product constant for water, K_w, increases with temperature. What will be the effect of lowering the temperature on the pH of pure water?

Concept Target

• Examine the relationship between K_w and pH of pure water.

Solution

When you lower the temperature of pure water, the value of K_w decreases. In pure water, the hydronium ion concentration equals the hydroxide ion concentration, so $K_w = [H_3O^+]^2$. When K_w decreases, the hydronium ion decreases, and the corresponding pH increases.

Conceptual Problem 16.20

You make solutions of ammonia and sodium hydroxide by adding the same moles of each solute to equal volumes of water. Which solution would you expect to have the higher pH?

Concept Target

• Underscore the meaning of a weak base versus that of a strong base.

Solution

Sodium hydroxide is a strong base, whereas ammonia is weak. As a strong base, NaOH exists in solution completely as ions, whereas NH_3 exists in solution as an equilibrium in which only part of the NH_3 has reacted to produce ions. Thus, a sodium hydroxide solution has a greater OH^- concentration than the same concentration solution of NH_3. At the same concentrations, the pH of the NaOH solution is greater (more basic) than the NH_3 solution.

Chapter 17

Acid-Base Equilibria

Concept Check 17.1

You have prepared dilute solutions of equal molar concentrations of $HC_2H_3O_2$ (acetic acid), HNO_2, HF, and HCN. Rank the solutions from the highest pH to the lowest pH.

Concept Target

• Underscore the relationship between K_a of an acid and the pH of solution.

Solution

You would probably guess that the pH's of the acid solutions depend on their respective K_a's: the larger the K_a, the greater the acidity or the lower the pH. We can put this on a firm basis by looking at the acid ionization equilibrium. An acid HA ionizes in water as follows:

$$HA(aq) + H_2O(l) \rightleftharpoons H_3O^+(aq) + A^-(aq)$$

The corresponding equilibrium constant K_a equals $[H_3O^+][A^-]/[HA]$. When you start with the same concentration of HA, the concentration of HA in solution is essentially the same for each acid. Also, $[H_3O^+] = [A^-]$. This means that K_a is proportional to $[H_3O^+]^2$. Or, $pH = -\log [H_3O^+]$ is proportional to $-\log K_a$. Therefore, the larger the K_a, the lower the pH. As an example, compare two acids, one with $K_a = 10^{-5}$ and the other with K_a equal to 10^{-4}. The corresponding $-\log K_a$ values are 5 and 4, respectively. The second acid (the one with the greater K_a) would have the lower pH. If you look at Table 17.1, the acid with the largest K_a of those listed in the problem statement is HF. So the ranking from lowest to highest pH is: $HCN < HC_2H_3O_2 < HNO_2$, HF.

Concept Check 17.2

Which of the following aqueous solutions has the highest pH and which has the lowest?
a. 0.1 M NH_3

b.   0.1 M NH_4Br
c.   0.1 M NaF
d.   0.1 M $NaCl$

Concept Target

- Use qualitative ideas of acid and base strengths of salt solutions.

Solution

Ammonia, NH_3, is a weak base; the other compounds are salts. You can decide the acidity or basicity of salt solutions by noting whether the corresponding acid and base are strong or weak. For example, NH_4Br is the salt of a weak base (NH_3) and a strong acid (HBr), so the salt is acidic. Similarly, NaF is basic (it is the salt of a strong base, NaOH, and a weak acid, HF). NaCl is neutral. This means that two of the solutions are basic (NH_3 and NaF), one solution is neutral (NaCl), and the other is acidic (NH_4Br). Although a salt might be as basic as NH_3, this occurs only when the acid from which the salt formed is quite weak (for example, NaCN is quite basic). So the solution with highest pH is 0.1 M NH_3 (a), and the solution of lowest pH is 0.1 M NH_4Br (b).

Concept Check 17.3

You add 1.5 mL of 1 M HCl to each of the following solutions. Which one will show the least change of pH?
a.   15 mL of 0.1 M NaOH
b.   15 mL of 0.1 M $HC_2H_3O_2$
c.   30 mL of 0.1 M NaOH and 30 mL of 0.1 M $HC_2H_3O_2$
d.   30 mL of 0.1 M NaOH and 60 mL of 0.1 M $HC_2H_3O_2$

Concept Target

- Recognize a buffer solution when given the solution composition.

Solution

The amount of any substance in the solution is proportional to the volume times molarity. Let's look at each solution.
a.   The 1.5 mL of 1 M HCl just neutralizes 15 mL of 0.1 M NaOH, giving a solution of NaCl. Thus, the pH of the original solution changes from very basic to neutral (pH = 7) after the addition of HCl.
b.   The solution of acetic acid changes from weakly acidic to strongly acidic with the addition of HCl.

c. Equal amounts of NaOH and $HC_2H_3O_2$ in the original solution exactly neutralize each other to produce the salt $NaC_2H_3O_2$, which is slightly basic. The addition of the strong acid HCl gives an acidic solution.

d. The 30 mL of NaOH reacts with 30 mL of $HC_2H_3O_2$ to give the salt $NaC_2H_3O_2$, leaving an equal amount of the corresponding acid $HC_2H_3O_2$. The result is a buffer solution. The addition of HCl to the buffer does not change the pH appreciably. (This is true as long as the amount of added acid does not overwhelm the capacity of the buffer. In this case, it does not. The amount of acid and base conjugates in the buffer are twice the amount of added acid.)

Only d. does not change appreciably in pH.

Conceptual Problem 17.17

You have 0.10-mol samples of three acids identified simply as HX, HY, and HZ. For each acid, you make up a 0.10 M solution by adding sufficient water to each of the acid samples. When you measure the pH of these samples, you find that the pH of HX is greater than the pH of HY, which in turn is greater than the pH of HZ.

a. Which of the acids is the least ionized in its solution?

b. Which acid has the largest K_a?

Concept Target

• Understand the relationship between the concentration of an acid, the degree of ionization of an acid and K_a.

Solution

a. The greater the pH of the solution, the less H_3O^+ present in solution. When comparing the three acid solutions of equal concentration, the HX produced the least amount of H_3O^+ in solution (highest pH); therefore, it must be the least ionized.

b. When comparing the three acids, HZ was the most ionized in solution (lowest pH) producing the greatest concentration of H_3O^+. Therefore, HZ must be the strongest acid with the largest K_a.

Conceptual Problem 17.18

What reaction occurs when each of the following is dissolved in water?

a. HF

b. NaF

c. $C_6H_5NH_2$

d. $C_6H_5NH_3Cl$

Concept Target

• Recognize proton-transfer reactions.

Solution

a. When HF is dissolved in water, the F^- hydrolyzes. The reaction is
$HF(aq) + H_2O(l) \rightleftharpoons H_3O^+(aq) + F^-(aq)$.

b. When NaF is dissolved in water, F^- hydrolyzes. The reaction is
$F^-(aq) + H_2O(l) \rightleftharpoons HF(aq) + OH^-(aq)$.

c. When $C_6H_5NH_2$ is dissolved in water, the following reaction occurs:
$C_6H_5NH_2(aq) + H_2O(l) \rightleftharpoons C_6H_5NH_3^+(aq) + OH^-(aq)$.

d. When $C_6H_5NH_3Cl$ is dissolved in water, $C_6H_5NH_3^+$ hydrolyzes. The reaction is
$C_6H_5NH_3^+(aq) + H_2O(l) \rightleftharpoons C_6H_5NH_2(aq) + OH^-(aq)$.

Conceptual Problem 17.19

You have the following solutions, all of the same molar concentrations: KBr, HBr, CH_3NH_2, and NH_4Cl. Rank them from the lowest to the highest hydroxide-ion concentrations.

Concept Target

• Recognize the approximate acidity or basicity of solutions.

Solution

The KBr solution is made from a neutral salt. Thus, it should have a pH of 7.0. The HBr solution is a strong acid, with pH < 7.0. The CH_3NH_2 solution is a weak base, with pH > 7.0. The NH_4Cl solution is made from a salt with a weak acid cation, NH_4^+, with pH < 7.0. Therefore, the ranking from lowest to highest hydroxide concentrations is: HBr < NH_4Cl < KBr < CH_3NH_2.

Conceptual Problem 17.20

Rantidine is a nitrogen base that is used to control stomach acidity by suppressing the stomach's production of hydrochloric acid. The compound is present in Zantac® as the chloride salt (rantidinium chloride; also called rantidine hydrochloride). Do you expect a solution of rantidine hydrochloride to be acidic, basic, or neutral? Explain by means of a general chemical equation.

Concept Target

• Understand that the cation corresponding to a base is the conjugate acid.

Solution

A solution of rantidine hydrochloride should be acidic. Let Ran represent rantidine. Then RanHCl is the chloride salt, rantidinium chloride. The cation rantidinium, $RanH^+$, should hydrolyze according to the following equation.

$$RanH^+(aq) + H_2O(l) \rightleftharpoons Ran(aq) + H_3O^+(aq)$$

Conceptual Problem 17.21

A chemist prepares dilute solutions of equal molar concentrations of NH_3, NH_4Br, NaF, and $NaCl$. Rank these solutions from highest to lowest pH.

Concept Target

• Compare the pH of various solutions.

Solution

NH_3 is a weak base; therefore, the solution would have pH >7. The NH_4Br solution is made from a salt with a weak acid cation, NH_4^+, with pH < 7.0. The NaF solution is made from a salt with a basic anion, F^-, with a pH > 7. The NaCl solution is made from a neutral salt; therefore, it should have a pH = 7. In order to rank the relative strengths of the F^- and NH_3, you should estimate K_b for each. Using data from Appendices E and F of your text, K_b for NH_3 is about 10^{-5} and about 10^{-10} for F^-, which indicates NH_3 is a stronger base than F^-. Therefore, the ranking from highest to lowest pH is: $NH_3 > F^- > NaCl > NH_4Br$.

Conceptual Problem 17.22

You want to prepare a buffer solution that has a pH equal to the pK_a of the acid component of the buffer. If you have 100 mL of a 0.10 M solution of the acid HA, what volume and concentration of NaA solution could you use in order to prepare the buffer?

Concept Target

• Develop an understanding of the conditions where the pK_a = pH of buffer solutions.

Solution

To prepare a buffer solution that has a pH equal to the pK_a of the acid, you need equal amounts of acid and conjugate base in the solution. The easiest way to make the buffer is to mix equal volumes of equal molar solutions of HA and NaA. Thus, mix 100 mL of 0.10 M NaA with the 100 mL of 0.10 M HA to prepare the buffer.

Conceptual Problem 17.23

A friend of yours has performed three titrations: strong acid with a strong base, weak acid with a strong base, and a weak base with a strong acid. He hands you the three titration curves, saying he has forgotten which is which. What attributes of the curves would you look at to correctly identify each curve?

Concept Target

• Learn the distinguishing features of different types of titrations.

Solution

Of the three titrations, the weak base-strong acid titration is easiest to identify. The attributes to look for are a high pH at the start, and a gradual decrease in pH as the titration proceeds. The other two titrations both start with low pH and increase in pH as the titration proceeds. Of the two, the strong acid-strong base titration should show a more rapid increase in pH before the endpoint is reached, and a well-defined, sharp equivalence point at pH 7. The weak acid-strong base titration will show a more gradual change in pH before the equivalence point, which should be less sharply defined, and occur at a pH above 7.

Conceptual Problem 17.24

You are given the following acid-base titration data, where each point on the graph represents the pH after adding a given volume of titrant (substance being added during the titration).

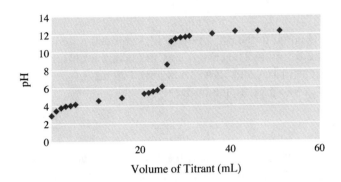

a. What substance is being titrated, a strong acid, strong base, weak acid, or weak base?
b. What is the pH at the equivalence point of the titration?
c. What indicator might you use to perform this titration? Explain.

Concept Target

• Learn the distinguishing features of different types of titration curves.

Solution

There are several features of the curve that indicate the type of titration that is being performed.

a. First, since the pH of the solution is less than 7 at the beginning of the titration, you can conclude that an acid is being titrated with a base.
b. The pH of the equivalence point is about 8.5. Therefore, the acid that is being titrated is a weak acid.
c. You need to pick an indicator that changes color in the pH range of about 7-10. Therefore, Thymol blue or phenolphthalein would work fine for this titration.

Chapter 18

Solubility and Complex-Ion Equilibria

Concept Check 18.1

Lead compounds have been used as pigments to make paint. Several factors are important to consider in contemplating the use of a compound as a paint pigment. Toxicity is one factor, and because of their toxicity, lead compounds are used less often today. Solubility of a compound is another factor; normally, the compound should be insoluble. Comparing solubility product constants, which of these compounds of lead is least soluble, $PbCrO_4$, $PbSO_4$, or PbS?

Concept Target

• Understand the qualitative relationship between K_{sp} and solubility.

Solution

Solubility and K_{sp} are related, although not directly. You can compare K_{sp}'s for a series of salts, however, if they have the same number of cations and anions in each of their formulas. (In that case, K_{sp} and solubility are related in the same way for each salt.) In this problem, each of the lead(II) compounds has one Pb^{2+} cation and one anion, so you can compare the K_{sp}'s directly. Lead(II) sulfide has the smallest K_{sp} and, therefore, is the least soluble of these lead(II) compounds.

Concept Check 18.2

Suppose you have equal volumes of saturated solutions of $NaNO_3$, Na_2SO_4, and PbS. Which solution would dissolve the most lead(II) sulfate, $PbSO_4$?

Concept Target

• Strengthen understanding of the common-ion effect.

Solution

Let's look at each compound in turn. $NaNO_3$ has no ion in common with $PbSO_4$, so it should have little effect on its solubility. Na_2SO_4 is a soluble compound and provides the common ion SO_4^{2-}, which would repress the solubility of $PbSO_4$. PbS has an ion in common with $PbSO_4$ (Pb^{2+}), but the compound is so insoluble that very little of the Pb^{2+} ion is available. Because of this, the solubility of $PbSO_4$ is little affected by the PbS. Therefore, only Na_2SO_4 appreciably affects the solubility of $PbSO_4$.

Concept Check 18.3

If you add a dilute acidic solution to a mixture containing magnesium oxalate and calcium oxalate, which of the two compounds is more likely to dissolve?

Concept Target

• Underscore the role of acidity in solubility of slightly soluble salts.

Solution

If you compare the K_{sp}'s for magnesium oxalate (8.5×10^{-5}) and calcium oxalate (2.3×10^{-9}), you can see that magnesium oxalate is much more soluble in water solution than the calcium salt (the K_{sp} is larger). This means that it provides a greater concentration of oxalate ion. In water solution, some of the magnesium oxalate dissolves, giving an oxalate ion concentration that tends to repress the dissolution of calcium oxalate (common ion effect). The addition of acid tends to remove oxalate ion, but this is replenished by the dissolution of more magnesium oxalate. Therefore, you would expect the magnesium oxalate to be more likely to dissolve.

Conceptual Problem 18.11

Which compound in each of the following pairs of compounds is the more soluble one?
a.   silver chloride or silver iodide
b.   magnesium hydroxide or copper(II) hydroxide

Concept Target

• Understand the qualitative relationship between K_{sp} and solubility.

Solution

a. Since both compounds contain the same number of ions, the salt with the larger K_{sp} value will be more soluble. The K_{sp} for AgCl is 1.8×10^{-10}, while for AgI, it is 8.3×10^{-17}. Therefore, silver chloride (AgCl) is more soluble.

b. Since both compounds contain the same number of ions, the salt with the larger K_{sp} value will be more soluble. The K_{sp} for $Mg(OH)_2$ is 1.8×10^{-11}, while for $Cu(OH)_2$, it is 2.6×10^{-19}. Therefore, magnesium hydroxide ($Mg(OH)_2$) is more soluble.

Conceptual Problem 18.12

You are given two mineral samples: halite, which is NaCl, and fluorite, which is CaF_2. Describe a simple test you could use to discover which mineral is fluorite.

Concept Target

• Develop a sense of what is soluble and what is insoluble (review of solubility rules).

Solution

Dissolve a sample of each mineral in water in separate containers. NaCl is soluble in water, and CaF_2 is insoluble in water ($K_{sp} = 3.4 \times 10^{-11}$), so the sample that dissolves is NaCl, while the sample that doesn't dissolve is CaF_2.

Conceptual Problem 18.13

You are given a saturated solution of lead(II) chloride. Which one of the following solutions would be most effective in yielding a precipitate when added to the lead(II) chloride solution?

a. 0.1 M NaCl(aq)
b. 0.1 M Na_2SO_4(aq)
c. saturated PbS(aq)

Concept Target

• Understand the common-ion effect.

Solution

0.1 M NaCl will be most effective in causing a precipitate from a saturated solution of $PbCl_2$, since it is completely soluble, and contains the common ion Cl^-. Therefore, the answer is a.

Conceptual Problem 18.14

Which of the following pictures best represents a solution made by adding 10 g of silver chloride, AgCl, to a liter of water? In these pictures, the dark spheres represent Ag^+ and the light spheres represent chloride ions. For clarity, water molecules are not shown.

Concept Target

• Develop a molecular level understanding of a slightly soluble ionic compound in water.

Solution

The beaker on the left depicts all of the AgCl as individual formula units in solution. This implies that AgCl is a soluble nonelectrolyte, which is not the case since AgCl is a slightly soluble ionic compound. The center beaker depicts AgCl as a soluble ionic compound that completely dissolves in solution leaving no AgCl(s). Once again, this cannot be correct since AgCl is slightly soluble. The beaker on the right indicates that there are ions of Ag^+ and Cl^- present in the solution along with solid AgCl. This is consistent with AgCl being a slightly soluble ionic compound.

Conceptual Problem 18.15

Which of the following pictures best represents an unsaturated solution of sodium chloride, NaCl? In these pictures, the dark spheres represent Na^+ ions and the light spheres represent chloride ions. For clarity, water molecules are not shown.

Concept Target

• Develop a molecular level understanding of a soluble ionic compound in water.

Solution

The beaker on the left depicts all of the NaCl as individual formula units in solution. This implies that NaCl is a soluble nonelectrolyte, which is not the case since NaCl is a very soluble ionic compound that produces ions in solution. The center beaker depicts NaCl as a soluble ionic compound that completely dissolves in solution producing only $Na^+(aq)$ and $Cl^-(aq)$. This must be correct since NaCl is a very soluble ionic compound. The beaker on the right indicates that there are ions of Na^+ and Cl^- present in the solution along with solid NaCl. The presence of solid is not consistent with NaCl being a very soluble ionic compound.

Conceptual Problem 18.16

When ammonia is first added to a solution of copper(II) nitrate, a pale blue precipitate of copper(II) hydroxide forms. As more ammonia is added, however, this precipitate dissolves. Describe what is happening.

Concept Target

• Examine the formation of hydroxo complex ions.

Solution

When first added to a solution of copper(II) nitrate, the ammonia causes a pale blue precipitate of $Cu(OH)_2$ to form ($K_{sp} = 2.6$ x 10^{-19}). After enough ammonia has been added, however, the complex ion $Cu(NH_3)_4^{2+}$ forms ($K_f = 4.8$ x 10^{12}), and the precipitate dissolves.

Conceptual Problem 18.17

You are given a solution of the ions Mg^{2+}, Ca^{2+}, and Ba^{2+}. Devise a scheme to separate these ions using sodium sulfate. Note that magnesium sulfate is soluble.

Concept Target

• Reinforce the concept of fractional precipitation.

Solution

Add just enough Na_2SO_4 to precipitate all of the Ba^{2+}; filter off the $BaSO_4$; add more Na_2SO_4 to precipitate all of the Ca^{2+}; filter off the $CaSO_4$; Mg^{2+} remains in the solution.

Conceptual Problem 18.18

You add dilute hydrochloric acid to a solution containing a metal ion. No precipitate forms. After the acidity is adjusted to 0.3 M hydronium ion, you bubble hydrogen sulfide into the solution. Again no precipitate forms. Is it possible that the original solution contained silver ion? Could it have contained copper(II) ion?

Concept Target

• Understand the concepts involved in qualitative analysis.

Solution

When a precipitate fails to form when HCl is added to the solution, this indicates that no silver ion is present. When a precipitate fails to form when the solution is acidified and H_2S is added, this indicates that no copper(II) ion is present in the solution.

Chapter 19

Thermodynamics and Equilibrium

Concept Check 19.1

You have a sample of 1.0 mg of solid iodine at room temperature. Later, you notice that the iodine has sublimed (passed into the vapor state). What can you say about the change of entropy of the iodine?

Concept Target

• Practice using qualitative rules regarding the entropy change in a process.

Solution

The process is $I_2(s) \rightarrow I_2(g)$. The iodine atoms have gone from a state of some order (crystalline iodine) to one that is more disordered (gas). The entropy will have increased.

Concept Check 19.2

Consider the reaction of nitrogen, N_2, and oxygen, O_2, to form nitric oxide, NO:
    $N_2(g) + O_2(g) \rightarrow 2NO(g)$.
From the standard free energy of formation of NO, what can you say about this reaction?

Concept Target

• Understand qualitatively the relation between standard free energy and equilibrium composition of a reaction mixture.

Solution

The standard free energy of formation of NO(g) is 86.60 kJ/mol. That this is a rather large positive value means that the equilibrium constant is small. At equilibrium, the NO concentration is low.

Concept Check 19.3

The following reaction is spontaneous in the direction given.

$$A(g) + B(g) \rightarrow C(g) + D(g)$$

Suppose you are given a vessel containing an equilibrium mixture of A, B, C, and D, and you increase the concentration of C by increasing its partial pressure.

a.   How is the value of $\Delta G°$ affected by the addition of C to the vessel?
b.   How is the value of ΔG affected by the addition of C to the vessel?

Concept Target

• Explore the relationship between $\Delta G°$ and ΔG.

Solution

a.   The standard free-energy change $\Delta G°$ is independent of concentration, so it will not change.
b.   The relationship between ΔG and $\Delta G°$ is given by $\Delta G = \Delta G° + RT \ln Q$, where Q is the reaction quotient. If the concentration of C is increased, this causes the value of Q to increase, and in turn ΔG to increase.

Conceptual Problem 19.17

For each of the following statements, indicate whether it is true or false.

a.   A spontaneous reaction always releases heat.
b.   A spontaneous reaction is always a fast reaction.
c.   The entropy of a system always increases for a spontaneous change.
d.   The entropy of a system and its surroundings always increases for a spontaneous change.
e.   The energy of a system always increases for a spontaneous change.

Concept Target

• Understand the conditions for a spontaneous change.

Solution

a. False. The enthalpy change (heat of reaction) has no direct relation to spontaneity.
b. False. The rate of a reaction has nothing to do with the spontaneity (thermodynamics) of a reaction.
c. False. The entropy may increase or decrease during a spontaneous reaction.
d. True. The entropy of the system plus surroundings always increases during a spontaneous change.
e. False. The energy may increase or decrease during a spontaneous reaction.

Conceptual Problem 19.18

Which of the following are spontaneous processes?
a. A cube of sugar dissolves in a cup of hot tea.
b. A rusty crowbar turns shiny.
c. Butane from a lighter burns in air.
d. A clock pendulum, initially stopped, begins swinging.
e. Hydrogen and oxygen gases bubble out from a glass of pure water.

Concept Target

• Gain a sense of spontaneous and nonspontaneous processes.

Solution

a. Spontaneous. Sugar dissolves spontaneously in hot water.
b. Nonspontaneous. Rust does not spontaneously change to iron; rather iron spontaneously rusts in air.
c. Spontaneous. The burning of butane in air is a spontaneous reaction.
d. Nonspontaneous. A pendulum once stopped will not spontaneously begin moving again.
e. Nonspontaneous. Water will not spontaneously decompose into its elements.

Conceptual Problem 19.19

For each of the following series of pairs, indicate which one of each pair has the greater quantity of entropy.
a. 1.0 mol of carbon dioxide gas at 20°C, 1 atm, or 2.0 mol of carbon dioxide gas at 20°C, 1 atm.
b. 1.0 mol of butane liquid at 20°C, 10 atm, or 1.0 mol of butane gas at 20°C, 10 atm.
c. 1.0 mol of solid carbon dioxide at -80°C, 1 atm, or 1.0 mol of solid carbon dioxide at -90°C, 1 atm.

d.   25 g of solid bromine at -7°C, 1 atm, or 25 g of bromine vapor at -7°C, 1 atm.

Concept Target

• Understand which state of a substance has the greater entropy.

Solution

a.   The 2.0 mol of CO_2 at 20°C and 1 atm has a greater entropy; two moles of substance has more entropy than one mole.
b.   The 1.0 mol of butane gas at 20°C and 10 atm has the higher entropy; the gaseous state has more entropy than the liquid state of the same substance under the same conditions.
c.   The 1.0 mol of $CO_2(s)$ at -80°C and 1 atm has the higher entropy; a solid substance has more entropy at the higher temperature.
d.   The 25 g of bromine vapor at -7°C and 12 atm has the higher entropy; the gaseous (vapor) state has more entropy than the liquid state of the same substance under the same conditions.

Conceptual Problem 19.20

Predict the sign of the entropy change for each of the following processes.
a.   A drop of food coloring diffuses throughout a glass of water.
b.   A tree leafs out in the spring.
c.   Flowers wilt and stems decompose in the fall.
d.   A lake freezes over in the winter.
e.   Rainwater on the pavement evaporates.

Concept Target

• Understand that when there is an increase in order entropy decreases.

Solution

a.   Entropy increases; ΔS is positive; there is an increase of disorder when the food coloring disperses throughout the water.
b.   Entropy increases; ΔS is negative; as a tree leafs out, order increases, and entropy decreases.
c.   Entropy increases; ΔS is positive; as flowers wilt and stems decompose, order decreases and entropy increases.
d.   Entropy decreases; ΔS is negative; as a liquid changes to solid, there is an increase of order and a decrease of entropy.

e. Entropy increases; ΔS is positive; as a liquid changes to vapor, there is a decrease of order and the entropy increases.

Conceptual Problem 19.21

Here is a simple experiment. Take a rubber band and stretch it. (Is this a spontaneous process? How does the Gibbs free energy change?) Place the rubber band against your lips; note how warm the rubber band has become. (How does the enthalpy change?) According to polymer chemists, the rubber band consists of long, coiled molecules. On stretching the rubber band, these long molecules uncoil and align themselves in a more ordered state. Show how the experiment given here is in accord with this molecular view of the rubber band.

Concept Target

• Gain a sense of the signs of thermodynamic quantities for a process.

Solution

The process is not spontaneous; you have to stretch the rubber band (the opposite process, a stretched rubber band snapping to its normal shorter shape, is spontaneous). Thus, ΔG is positive. The fact that the stretched rubber band feels warm means that ΔH is negative (exothermic). Note that $\Delta G = \Delta H - T\Delta S$; so $\Delta S = -(\Delta G - \Delta H)/T$. This implies that ΔS is negative, which is consistent with an increase in order.

Conceptual Problem 19.22

Describe how you would expect the spontaneity ($\Delta G°$) for each of the following reactions to behave with a change of the temperature.
a. Phosgene, $COCl_2$, the starting material for the preparation of polyurethane plastics, decomposes as follows:
    $COCl_2 \rightarrow CO(g) + Cl_2(g)$.
b. Chlorine adds to ethylene to produce dichloroethane, a solvent:
    $Cl_2(g) + C_2H_4(g) \rightarrow C_2H_4Cl_2(l)$.

Concept Target

• Understand qualitatively how entropy and enthalpy are affected by bond formation and bond breaking, and how these thermodynamic quantities affect the spontaneity of a reaction with temperature.

Solution

a. $\Delta H°$ is positive for this reaction (you need energy to break bonds), and $\Delta S°$ is also positive (breaking a molecule into two increases disorder and entropy). Referring to Table 19.3, you see that the reaction is nonspontaneous at low T and spontaneous at high T. Therefore, the spontaneity ($\Delta G°$) of the reaction increases with temperature.

b. $\Delta S°$ for this reaction is negative (two molecules form one, so there is an increase in order and a decrease in entropy), and $\Delta H°$ is also negative (energy is released when bonds form). Referring to Table 19.3, you see that the reaction is spontaneous at low T and nonspontaneous at high T. Therefore, the spontaneity ($\Delta G°$) of the reaction decreases with temperature.

Chapter 20

Electrochemistry

Concept Check 20.1

If you were to construct a wet cell and decided to replace the salt bridge with a piece of copper wire, would the cell produce sustainable current? Be sure to explain your answer.

Concept Target

• Emphasize the role and function of the salt bridge in a wet cell.

Solution

No sustainable current would flow. The wire does not contain mobile positively and negatively charged species that are necessary to balance the accumulation of charges in each of the half-cells.

Concept Check 20.2

Let's say that we define the reduction of I_2 to I^- ions, $I_2(s) + 2e^- \rightarrow 2I^-(aq)$, as the standard reduction reaction with $E° = 0.00$ V. We then construct a new standard reduction table based on this definition.

a. What would be the new standard reduction potential of H^+?
b. Would using a new standard reduction table change the <u>measured</u> value of a freshly prepared voltaic cell made from Cu and Zn (assume you have the appropriate solutions and equipment to construct the cell)?
c. Would the calculated voltage for the cell in part b. be different than if you were using the values presented in Table 20.1? Do the calculations to justify your answer.

Concept Target

• Illustrate the fact that the values of E_{cell} from the table of standard reduction potentials are measured against an arbitrary standard.

Solution

a.   Standard reduction potentials are measured against some arbitrarily chosen standard reference half-reaction. Only differences in potentials can be measured. A voltaic cell made from H_2 and I_2, and corresponding solutions, will have the same voltage, regardless of the choice of the reference cell. If the I_2/I^- half-reaction is assigned a value of $E^\circ = 0.00$ V, then the H_2/H^+ half-reaction must have a voltage of $E^\circ = -0.54$ V to keep the overall voltage the same.

b.   The voltage of a voltaic cell made from Cu and Zn, and corresponding solutions, will have the same measured voltage, regardless of the choice of the reference half-reaction.

c.   The calculated voltage is 1.10 V and is the same either way.

Concept Check 20.3

Consider the voltaic cell: $Fe(s)|Fe^{2+}(aq)||Cu^{2+}(aq)|Cu(s)$ being run under standard conditions.

a.   Is ΔG° positive or negative for this process?

b.   Change the concentrations from their standard values in such a way that E_{cell} is reduced. Write your answer using the shorthand notation (Section 20.3).

Concept Target

• Qualitatively explore how cell concentrations affect ΔG° and E_{cell}.

Solution

a.   Using the standard reduction potentials in Table 20.1, you see that the voltage for this cell is positive, suggesting that ΔG° is negative.

b.   In order to reduce E_{cell}, you change the concentrations in a manner to increase the value of Q, where

$$Q = \frac{[Fe^{2+}]}{[Cu^{2+}]}$$

For example: $Fe(s)|Fe^{2+}\ (1.10\ M)||Cu^{2+}(0.50\ M)|Cu(s)$.

Concept Check 20.4

Keeping in mind that seawater contains a number of ions, explain why seawater corrodes iron much faster than fresh water.

Concept Target

• Use electrochemical information from the table of standard reduction potentials to explain corrosion.

Solution

Many of the ions contained in seawater have very high reduction potentials – higher than $Fe(s)$. This means that spontaneous electrochemical reactions will occur with the $Fe(s)$ causing the iron to form ions and go into solution, while at the same time, the ions in the sea are reduced and plate out on the surface of the iron.

Conceptual Problem 20.19

Keeping in mind the fact that aqueous Cu^{2+} is blue colored and aqueous Zn^{2+} is colorless, predict what you would observe over a several day period if you performed the following experiments.
a. A strip of Zn is placed into a beaker containing aqueous Zn^{2+}.
b. A strip of Cu is placed into a beaker containing aqueous Cu^{2+}.
c. A strip of Zn is placed into a beaker containing aqueous Cu^{2+}.
d. A strip of Cu is placed into a beaker containing aqueous Zn^{2+}.

Concept Target

• Use the information presented in the table of standard reduction potentials to predict the outcomes of chemical reactions.

Solution

a. Since there is no species present to donate or accept electrons other than zinc, you would expect no change.
b. Since there is no species present to donate or accept electrons other than copper, you would expect no change.
c. According to the table of standard reduction potentials, the Cu^{2+} would undergo reduction and the Zn would undergo oxidation. You would expect the Zn strip to dissolve as it becomes Zn^{2+}, the blue color of the solution to fade as the Cu^{2+} becomes Cu, and the formation of the solid copper precipitate.

d. According to the table of standard reduction potentials, since Zn^{2+} cannot oxidize Cu,
    you would expect no change.

Conceptual Problem 20.20

You are working at a plant that manufactures batteries. A client comes to you and asks for a
6.0-V battery that is made from silver and cadmium. Assuming that you are running the
battery under standard conditions, how should it be constructed?

Concept Target

• Emphasize that to obtain batteries of a higher voltage than provided by a single cell, several
  cells can be placed in series.

Solution

You could construct the battery by hooking together, in series, five individual cells, each
with $E° = 1.20$ V.

Conceptual Problem 20.21

The composition of the hull of a submarine is mostly iron. Pieces of zinc, called "zincs", are
placed in contact with the hull throughout the inside of the submarine. Why is this done?

Concept Target

• Apply information from the table of standard reduction potentials and illustrate an
  important use of cathodic protection.

Solution

The Zn is a sacrificial electrode, keeping the hull from undergoing oxidation by the dissolved
ions in sea water. Zn works because it is more easily oxidized than Fe.

Conceptual Problem 20.22

You place a battery in a flashlight in which all of the electrochemical reactions have reached
equilibrium. What do you expect to observe when you turn the flashlight on? Explain your
answer.

Concept Target

- Connect the concept of equilibrium in an electrochemical cell with a commonly observed event (a dead battery).

Solution

When an electrochemical reaction reaches equilibrium, $E_{cell} = 0$, which means no current will flow and nothing will happen when you turn on the flashlight. This is typically what has occurred when you have a dead battery.

Conceptual Problem 20.23

The difference between a "heavy-duty" and a regular zinc-carbon battery is that the zinc can in the heavy-duty battery is thicker walled. What makes this battery heavy-duty in terms of output?

Concept Target

- Illustrate the relationship between the quantity of reactants in an electrochemical cell and duration of the oxidation-reduction reaction.

Solution

Since there is more zinc present, the oxidation-reduction reactions in the battery will run for a longer period of time. This assumes that the zinc is the limiting reactant.

Conceptual Problem 20.24

From an electrochemical standpoint, what metal, other than zinc, would be a reasonable candidate to coat a piece of iron to prevent corrosion (oxidation)?

Concept Target

- Establish the relationship between the position of an element in the table of standard reduction potentials and its reactivity with other elements.

Solution

Any metal that has a more negative standard reduction potential (Mg, Al, etc.) could be used, keeping in mind that group IA metals that fall into this category are too reactive to be of practical use.

Conceptual Problem 20.25

Household bleach is a solution containing NaClO, which is a very strong oxidizing agent. Come up with a brief explanation as to how bleach is able to "whiten" stains in clothing.

Concept Target

• Make the connection between electrochemistry and the function of a commonly used household product (bleach).

Solution

A stain is visible in clothing because it is absorbing only a portion of the visible light striking the material (which means some light is reflected off the clothes, giving the stain color). When you bleach the stain, the compounds that make up the stain are oxidized. This oxidation often changes the electronic properties of the stain to make it absorb no colors in the visible region, thus reflecting back all wavelengths of visible light, making the stain disappear.

Conceptual Problem 20.26

The development of lightweight batteries is an ongoing research effort combining many of the physical sciences. You are a member of an engineering team trying to develop a lightweight battery that will effectively react with $O_2(g)$ from the atmosphere as the oxidizing agent. A reducing agent must be chosen for this battery that will be lightweight, have nontoxic products, and react spontaneously with oxygen. Using data from Appendix I, suggest a likely reducing agent, being sure the above conditions are met. Are there any drawbacks to your selection?

Concept Target

• Use electrochemical information to design a battery that has specific design and engineering requirements.

Solution

The most important consideration in battery design is the spontaneity of the cell reaction, since this determines the cell voltage. The half-reactions for the reduction of oxygen gas given in Appendix I are:

$$O_2 + 2H_2O + 4e^- \rightarrow 4OH^- \qquad E° = +0.40 \text{ V}$$
$$O_2 + 4H^+ + 4e^- \rightarrow 2H_2O \qquad E° = +1.23 \text{ V}$$

These represent the reduction of oxygen under basic and under acidic conditions. Thus, in basic solution, we would require an oxidation half-reaction (anode half-reaction) with a potential greater than -0.40 V to obtain a spontaneous reaction.

The density of the reducing agent is the next most important consideration. Hydrogen, the element with the lowest density, should certainly be considered. It does have some drawbacks, however. Because it is a gas, some method of storage needs to be developed. Liquid storage and metal hydride storage have been investigated. Storage of liquid hydrogen presents problems of safety and weight. Storage of hydrogen as the hydride is used in nickel-hydride cells presently available for portable computers, cellular phones, etc. (These batteries are rechargeable, but do not use oxygen.) The metal hydride obviously adds weight to the battery. Other elements you might consider are Na, Li, Al, K, Ca, and Zn, which have favorable power-to-mass ratios. Lithium and sodium might present some disposal problems since they are very reactive metals. Batteries using aluminum with atmospheric oxygen are available.

Chapter 21

Nuclear Chemistry

Concept Check 21.1

You have two samples of water each made up of different isotopes of hydrogen: one contains $^1_1 H_2O$ and the other, $^3_1 H_2O$.

a. Would you expect these two water samples to be chemically similar?
b. Would you expect these two samples to be physically the same?
c. Which one of these water samples would you expect to be radioactive?

Concept Target

• Highlight the relationships between the isotopes of an element and its chemical and physical properties.

Solution

a. Yes. Isotopes have similar chemical properties.
b. No, since the $^3_1 H_2O$ molecule is more massive than $^1_1 H_2O$.
c. $^3_1 H_2O$ should be radioactive.

Concept Check 21.2

Say you are internally exposed to 10 rads of α, β, and γ radiation. Which form of radiation will cause the greatest biological damage?

Concept Target

• Underscore that the amount of biological damage caused by radiation is a function of both the amount and type of radiation.

Solution

For the same radiation dosage (10 rads) the form of radiation with the highest RBE will cause the greatest biological damage. Therefore, the α particle will cause the most damage, since it has the highest RBE (10).

Concept Check 21.3

Why do you think that carbon-14 dating is limited to about 50,000 years?

Concept Target

• Illustrate that very little radioactive material remains after 10 half-lives.

Solution

After 50,000 years, enough half-lives have passed (about 10) so there would be almost no carbon-14 present to detect and measure (about 0.1% would be left).

Conceptual Problem 21.19

When considering the lifetime of a radioactive species, a general rule of thumb is that after 10 half-lives have passed, the amount of radioactive material left in the sample is negligible. The disposal of some radioactive materials is based on this rule.
a. What percentage of the original material is left after 10 half-lives?
b. When would it be a bad idea to apply this rule?

Concept Target

• Demonstrate that in some cases even a small percentage of an original amount of radioactive material can still be significant.

Solution

a. After 10 half-lives have passed, the percentage of the original material that is left is

$$\left(\frac{1}{2}\right)^{10} \times 100\% = 0.0976 \cong 0.1\%.$$

b. If you had a large quantity of material, 0.1% still would be a significant quantity. Also, if the material were particularly toxic in addition to being radioactive, even small amounts would be a problem.

Conceptual Problem 21.20

Identify the following reactions as fission, fusion, or a transmutation, or radioactive decay.

a. $4\,^1_1 H \rightarrow \,^4_2 He + 2\,^0_1 e$

b. $^{14}_6 C \rightarrow \,^{14}_7 N + \,^0_{-1} e$

c. $^1_0 n + \,^{235}_{92} U \rightarrow \,^{140}_{56} Ba + \,^{93}_{36} Kr + 3\,^0_1 e$

d. $^{14}_7 N + \,^4_2 He \rightarrow \,^{17}_8 O + \,^1_1 H$

Concept Target

• Identify nuclear reaction types from nuclear equations.

Solution

a. fusion
b. radioactive decay
c. fission
d. transmutation

Conceptual Problem 21.21

Sodium has only one naturally occurring isotope, sodium-23. Using the data presented in Table 21.3, explain how the molecular weight of sodium is 22.98976 amu and not the sum of the masses of the protons, neutrons, and electrons.

Concept Target

• Show how the mass defect must be used to account for the fact that the mass of an isotope is not the sum of the masses of its constituent protons, neutrons, and electrons.

Solution

An atom of sodium-23 has 11 protons, 12 neutrons, and 11 electrons. Using the values in Table 21.3, the atomic mass of sodium-23 would be

(11)(1.00728 amu) + (12)(1.008665 amu) + (11)(0.000549 amu) = 23.190099 amu.

Since the observed atomic weight is 22.98976 amu, it is not the same as the sum of the masses of the protons, neutrons, and electrons. Some of the expected mass is in the form of energy: the mass defect.

Conceptual Problem 21.22

You have a mixture that contains 10 g of Pu-239 with a half-life of 2.4×10^4 years and 10 g of Np-239 with a half-life of 2.4 days. Estimate how much time must elapse before the quantity of radioactive material is reduced by 50%.

Concept Target

• Use half-life information to qualitatively predict the outcome of an experiment involving radioactive isotopes.

Solution

The total amount of radioactive material at the start is 10 g Pu-239 + 10 g Np-239 = 20 g total. In order to reduce this by 50%, the total mass must be reduced to 10 g. Since the half-life of Pu-239 (2.4×10^4 years) is long compared to the half-life of Np-239 (2.4 days), all of the Np-239 will have decomposed before any measurable amount of Pu-239 decays. This would require approximately 10 half-lives, or about 24 days.

Conceptual Problem 21.23

Come up with an explanation as to why α radiation is easily blocked by materials such as a piece of wood, where γ radiation easily passes through.

Concept Target

• Ensure a fundamental understanding of the basic atomic structure of matter and the properties of α and γ radiation.

Solution

The large, positively charged He nucleus that makes up alpha (α) radiation is unable to pass through the atoms that make up solid materials such as wood without coming into contact or being deflected by the nuclei. Gamma (γ) radiation, however, with its small wavelength and high energy, can pass through large amounts of material without interaction, just like x-rays can pass through skin and other soft tissue.

Conceptual Problem 21.24

You have an acquaintance who tells you that he is going to reduce his radiation exposure to zero. What examples could you present that would illustrate that this is an impossible goal.

Concept Target

• Underscore that radiation is a naturally occurring part of our world and, therefore, is not something that can be avoided.

Solution

Examples of elements and compounds that would be impossible to avoid include: radioactive ^{40}K that is in bananas and any food that contains potassium, H_2O that contains ^3_1H, CO_2 that contains carbon-14, and radon gas that comes from soil and rocks.

Conceptual Problem 21.25

In Chapter 7 (*A Chemist Looks At: Zapping Hamburger with Gamma Rays*) there is a discussion of how gamma radiation is used to kill bacteria in food. As indicated in the feature, there is concern on the part of some people that the irradiated food is radioactive. Why is this not the case? If you wanted to make the food radioactive, what would you have to do?

Concept Target

• Develop an understanding of how radiation interacts with matter and what makes something radioactive.

Solution

Irradiation of food atoms with gamma radiation does not result in the creation of radioactive elements; therefore, the food cannot become radioactive. The addition of a radioactive element directly to the meat would make the meat radioactive. Under the right conditions, nuclear bombardment could also lead to the production of radioactive elements.

Conceptual Problem 21.26

You have a pile of I-131 atoms with a half-life of 8 days. A portion of the solid I-131 is represented below. Can you predict how many half-lives will occur before the lighter colored (green in text) I-131 atom undergoes decay?

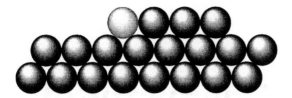

Concept Target

• Illustrate that the half-life of an isotope allows us to predict the behavior of a collection of atoms (bulk) but not the behavior of individual atoms.

Solution

No, half-life can only tell you the quantity of material that will undergo decay, not the identity of the individual atoms.

Chapter 22

Metallurgy and Chemistry of the Main Group Metals

Concept Check 22.1

As discussed in the text, Zr is used as cladding for nuclear fuel rods in power plants. Why don't you want to use a fuel rod cladding material that absorbs neutrons?

Concept Target

• Relate the properties of an element to a potential application.

Solution

The neutrons are what causes fission and a sustainable nuclear reaction. Therefore, an element that blocks the neutrons prevents a nuclear reaction.

Concept Check 22.2

Considering the fact that N_2 makes up about 80% of the atmosphere, why don't animals use the abundant N_2 instead of O_2 for biological reactions?

Concept Target

• Make the connection between the molecular structure of important atmospheric molecules and the function of these molecules in a biological system.

Solution

Given the high energy demands of animals to move and maintain body temperature, breaking the very strong triple bond of N_2 requires too much energy when compared to the lower energy double bond of O_2.

Concept Check 22.3

Why do we need such low temperatures to liquefy gasses such as nitrogen, oxygen, and He?

Concept Target

• Revisit the important relationships between molecular structure, intermolecular attractions, and the states of matter.

Solution

The only intermolecular forces in these materials are very weak van der Waals forces.

Conceptual Problem 22.59

When producing Coke, why is the coal heated in the absence of air? Write the chemical reaction for what would happen when it is heated in air.

Concept Target

• Predict the outcome of a chemical reaction.

Solution

If air is present, the oxygen in the air would react with the coal (undergo oxidation) by the following equation: $C(s) + O_2(g) \rightarrow CO_2(g)$.

Conceptual Problem 22.60

Even though hydrogen isn't a metal, why is it in group IA of most periodic tables?

Concept Target

• Reaffirm that the location of an element in the periodic table is based on the electronic structure of the element.

Solution

It has one valence electron like the other elements in group IA.

Conceptual Problem 22.61

What happens to the metallic character of the main-group elements as you move left to right across any row of the periodic table? What happens to the metallic character of the main-group elements as you move down a column (group)?

Concept Target

• Extract meaningful information about the elements from the periodic table.

Solution

The metallic character decreases from left to right and increases going down a column.

Conceptual Problem 22.62

Lithium hydroxide, like sodium hydroxide, becomes contaminated when exposed to air. What is the source of this contamination? What reactions take place?

Concept Target

• Make a connection between the chemistry of a well-described compound, NaOH, and another closely related compound.

Solution

Lithium hydroxide is contaminated by reaction with carbon dioxide absorbed from the air:
$$2LiOH(s) + CO_2(g) \rightarrow Li_2CO_3(s) + H_2O(l)$$
$$Li_2CO_3(s) + H_2O(l) + CO_2(g) \rightarrow 2LiHCO_3(s).$$

Conceptual Problem 22.63

Aluminum hydroxide is an amphoteric substance. What does this mean? Write equations to illustrate.

Concept Target

• Develop a clear understanding of the term amphoteric.

Solution

This means that aluminum hydroxide reacts with both acids and bases. For example,

$Al(OH)_3(s) + 3HCl(aq) \rightarrow 3H_2O(l) + AlCl_3(aq)$ and

$Al(OH)_3(s) + NaOH(aq) \rightarrow Na^+(aq) + Al(OH)_4^-(aq)$.

Conceptual Problem 22.64

Tin metal would not make a very good structural metal in cold climates. Why?

Concept Target

• Practice making connections between the chemical and physical properties of elements and their potential applications.

Solution

Below 13°C the stable white metallic allotrope undergoes a transition to the brittle powder allotrope called gray tin.

Conceptual Problem 22.65

Oxygen, like other second-period elements, is somewhat different from the other elements in its group. List some of these differences.

Concept Target

• Highlight the principal differences in electronic structure between oxygen and the other members of its group on the periodic table.

Solution

Oxygen is a very electronegative element, and its bonding involves only the s and p orbitals, in contrast to bonding using the d orbitals in sulfur, etc. Molecular oxygen is a reactive gas but forms mainly compounds in which its oxidation state is -2, compared to compounds of sulfur, etc., which exhibit positive oxidation states as well as the -2 state.

Conceptual Problem 22.66

Given the reaction $Cl_2(g) + 2KBr(aq) \rightarrow 2KCl(aq) + Br_2(aq)$ readily occurs, would you expect the reaction $I_2(s) + 2KCl(aq) \rightarrow 2KI(aq) + Cl_2(aq)$ to occur?

Concept Target

• Employ electrochemical information to predict the results of a chemical reaction.

Solution

Applying your chemical knowledge and consulting the table of standard reduction potentials (Appendix I), you would not expect I_2 to be a better oxidizing agent than Cl_2.

Conceptual Problem 22.67

Hydrogen chloride can be prepared by heating NaCl with concentrated sulfuric acid. Why is substituting NaBr for NaCl in this reaction not a satisfactory way to prepare HBr?

Concept Target

• Underscore the importance of oxidation and reduction potentials when trying to predict the outcome of many chemical reactions.

Solution

Consulting the table of standard reduction potentials (Appendix I), you find that HBr cannot be prepared by adding sulfuric acid to NaBr because the hot concentrated acid will oxidize the bromide ion to bromine.

Conceptual Problem 22.68

Do you expect an aqueous solution of sodium hypochlorite to be acidic, neutral, or basic? What about an aqueous solution of sodium perchlorate?

Concept Target

• Apply fundamental acid base concepts to a new problem.

Solution

An aqueous solution of sodium hypochlorite should be basic because HClO is a weak acid. A solution of sodium perchlorate should be neutral because $HClO_4$ is a strong acid and NaOH is a strong base.

Chapter 23

The Transition Elements

Concept Check 23.1

Another complex studied by Werner has a composition corresponding to the formula $PtCl_4 \cdot 2KCl$. From electrical-conductance measurements, he determined that each formula unit contained three ions. He also found that silver nitrate did not give a precipitate of AgCl with this complex. Write a formula for this complex that agrees with this information.

Concept Target

• Understand the experimental basis of the structural formulas of complexes.

Solution

Because silver nitrate does not precipitate AgCl from the complex, you conclude that chlorine is not present as free chloride ions; the Cl is presumably present as a complex. Potassium ion is likely present as K^+, which would account for two of the ions present in each formula unit. The other would be a complex ion of platinum with the six chlorine atoms (a complex of Pt with six Cl$^-$ ions). The charge on this complex ion must be -2 to counter the charges from the K^+ ions, so its formula is $PtCl_6^{2-}$. The formula of the complex then is $K_2[PtCl_6]$.

Concept Check 23.2

A complex has the composition $Co(NH_3)_4(H_2O)BrCl_2$. Conductance measurements show that there are three ions per formula unit, and precipitation of AgCl with silver nitrate shows that there are two Cl$^-$ ions not coordinated to cobalt. What is the structural formula of the compound? Write the structural formula of an isomer.

Concept Target

• Understand the basis of isomerism of complexes.

Solution

The addition of silver nitrate to the complex precipitates AgCl equivalent to two Cl⁻ ions per formula unit. Since each formula unit consists of three ions, the complex appears to consist of two Cl⁻ ions plus a complex ion with a charge of +2. The formula of the complex then would be $[Co(NH_3)_4(H_2O)Br]Cl_2$. The structural formula of a possible constitutional isomer is $[Co(NH_3)_4(H_2O)Cl]BrCl$.

Concept Check 23.3

a.   Which of the following molecular models of octahedral complexes are mirror images of the molecule X? Keep in mind that you can rotate the molecules when performing comparisons.
b.   Which complexes are optical isomers of molecule X?
c.   Identify the distinct geometric isomers of the complex X (Note that some of the models may represent the same molecule).

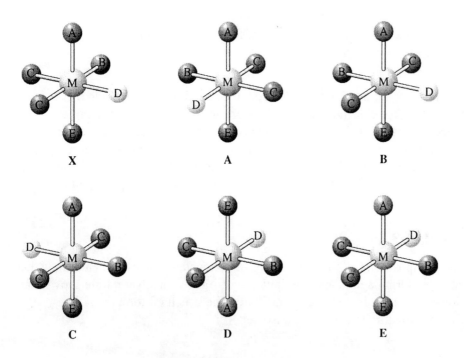

Concept Target

- Visualize complexes and recognize geometric and optical isomers based on molecular drawings.

Solution

a. If you were to place a mirror to the right of the complex X, complex A would directly represent what you would see in the mirror, so it is a mirror image. If you were to place a mirror behind complex X, complex D would also be a mirror image. Note that complexes B and C are exactly the same, and A and E are the same, in both cases, they only differ by rotation (spin one of the complexes in each pair 180° to see this). Therefore, A, D, and E are mirror images.
b. Optical isomers are nonsuperimposable mirror images of one another. The mirror images of molecule X, which are molecules A and D, are both superimposable mirror images, so no optical isomers of X are present.
c. To answer this part, you need to rotate each of the complexes to see if they have the same bonding arrangement in space with each of the ligands. Models A & E are the same complex, neither of which has the same bonding arrangement as complex X, so the complex that they represent is a geometric isomer. Models B & C represent the same complex that also has a different bonding arrangement than complex X, so the complex that they represent is also a geometric isomer. Complex D is the same complex as complex X, so it is not a geometric isomer.

Conceptual Problem 23.27

A cobalt complex whose composition corresponded to the formula $Co(NO_2)Cl \cdot 4NH_3$ gave an electrical conductance equivalent to two ions per formula unit. Excess silver nitrate solution immediately precipitated 1 mol AgCl per formula unit. Write a structural formula consistent with these results.

Concept Target

- Understand the experimental basis of the structural formulas of complexes.

Solution

Because one mole of chloride ion is precipitated per formula unit of the complex, the chlorine atoms must be present as chloride ion. All the other ligands are coordinated to the cobalt. An appropriate formula is $[Co(NH_3)_4(NO_2)_2]Cl$.

Conceptual Problem 23.28

For the following coordination compounds, identify the geometric isomer(s) of compound X.

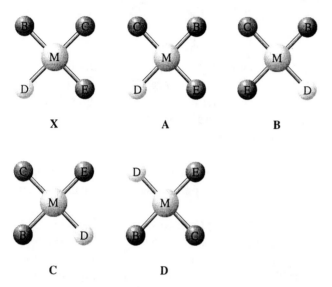

Concept Target

• Visualize and recognize geometric isomers from molecular drawings.

Solution

In order to solve this problem, you need to check each of the possible coordination compounds (A-D) to see if they are different than compound X. Compounds B, C, and D, through rotation and flipping over, all turn out to be the same as compound X. The only compound that is different is A, so it is the only geometric isomer of compound X.

Conceptual Problem 23.29

Describe step by step how the name potassium hexacyanoferrate(II) leads one to the structural formula $K_4[Fe(CN)_6]$.

Concept Target

• Emphasize the relationship between the name of a complex and its structure.

Solution

"Hexacyano" means that there are six CN$^-$ ligands bonded to the iron cation. The Roman numeral II means that the oxidation state of the iron cation is +2, so that the overall charge of the complex ion is -4. This requires four potassium ions to counterbalance the -4 charge.

Conceptual Problem 23.30

Compounds A and B are known to be stereoisomers of one another. Compound A has a violet color; compound B has a green color. Are they geometric or optical isomers?

Concept Target

• Realize that optical isomers must have the same physical properties.

Solution

Compounds A and B are geometric isomers because these isomers have different physical properties whereas optical isomers do not.

Conceptual Problem 23.31

A complex has a composition corresponding to the formula $CoBr_2Cl \cdot 4NH_3$. What is the structural formula if conductance measurements show two ions per formula unit? Silver nitrate solution gives an immediate precipitate of AgCl but no AgBr. Write the structural formula of an isomer.

Concept Target
• Understand the relationship between structure of a complex and possible isomers.

Solution

The given compound must consist of a chloride ion (which can be precipitated with $AgNO_3$ solution) and a $[Co(NH_3)_4Br_2]^+$ ion, giving $[Co(NH_3)_4Br_2]Cl$. The structural formula of a possible constitutional isomer is $[Co(NH_3)_4BrCl]Br$.

Conceptual Problem 23.32

For the complexes shown here, which complex anion would have the d electron distribution shown in the diagram below: MF_6^{3-}, $M(CN)_6^{3-}$, MF_6^{4-}, $M(CN)_6^{4-}$? Note that the neutral metal atom, M, in each complex is the same and has the ground state electron configuration $[Ar]4s^23d^6$.

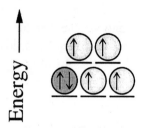

Concept Target

• Apply crystal field theory.

Solution

The d orbital energy level diagram contains six electrons. This indicates that the metal M must have the electron configuration $[Ar]3d^6$ and therefore a charge of 2+ (M^{2+}). The two complexes that have M in the 2+ oxidation state are MF_6^{4-} and $M(CN)_6^{4-}$. The d orbital energy level diagram is high spin indicating that there is a relatively small amount of crystal field splitting. According to the spectrochemical series, F^- is a less strongly bonding ligand with a relatively smaller crystal field splitting, which makes the MF_6^{4-} the most likely answer.

Chapter 24

Organic Chemistry

Concept Check 24.1

Given the model below where C atoms are labeled and H atoms are the small white balls:

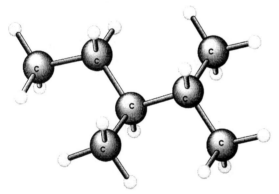

a. Write the molecular formula.
b. Write the condensed structural formula.

Concept Target

• Interpret the molecular model of a branched alkane.

Solution

a. C_7H_{16}

b.
$$
\begin{array}{c}
\quad\quad\quad CH_3 \\
\quad\quad\quad | \\
CH_3CHCHCH_2CH_3 \\
\quad\quad\quad | \\
\quad\quad\quad CH_3
\end{array}
$$

Concept Check 24.2

For a gasoline to function properly in an engine, it should not begin to burn before it is ignited by the spark plug. If it does, it makes the noise we think of as engine "knock." The octane-number scale rates the anti-knock characteristics of a gasoline. This linear scale is based on heptane, given an octane number of 0, and on 2,2,4-trimethylpentane (an octane constitutional isomer), given an octane number of 100. The higher the octane number, the better the anti-knock characteristics. If you had a barrel of heptane and a barrel of 2,2,4-trimethylpentane, how would you blend these compounds to come up with a 90 octane mixture?

Concept Target

• Illustrate that gasoline contains a mixture of hydrocarbons blended to make the desired octane.

Solution

Since, by definition, heptane is zero octane and 2,2,4-trimethylpentane is 100 octane, a mixture of 10% heptane and 90% 2,2,4-trimethylpentane would produce a 90 octane mixture.

Concept Check 24.3

Given the model below where C atoms are labeled and H atoms are the small white balls:

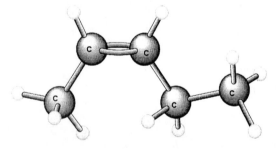

a. Write the molecular formula.
b. Write the condensed structural formula.
c. Write the IUPAC name.

Concept Target

• Interpret the molecular model of a geometric isomer.

Solution

a. C_5H_{10}

b.

$$\underset{CH_3}{\overset{H}{\diagdown}}C=C\underset{CH_2CH_3}{\overset{H}{\diagup}}$$

c. *cis*-2-pentene

Conceptual Problem 24.15

You are distilling a barrel of oil that contains the hydrocarbons listed in Table 24.4. You heat the contents of the barrel to 200°C.

a. What molecules will no longer be present in your sample of oil?
b. What molecules will be left in the barrel?
c. Provide an explanation for your answers in parts a. and b.
d. Which molecule would boil off at a lower temperature, hexane or 2,3-dimethylbutane?

Concept Target

• Explore the relationships between the structure and molecular weight of a hydrocarbon and its boiling point.

Solution

a. The molecules with carbon chains in the C_5-C_{11} range will no longer be present because they have all boiled off. Additionally, some of the heavily branched C_{12} chains will boil off.
b. The molecules with carbon chains greater than C_{11} will be left in the barrel because they boil at temperatures above 200°C (Keep in mind that the heavily branched C_{12} chains will have boiled off at or slightly below 200°C.)
c. Low molecular weight hydrocarbons have fewer polarizable electrons; therefore, they have weaker London forces than the longer chains and, as a result, boil at a lower temperature.
d. The more compact 2,3-dimethylbutane would boil at a lower temperature.

Conceptual Problem 24.16

A classmate tells you that the following compound has the name 3-propylhexane.

$$CH_2CH_2CH_3$$
$$|$$
$$CH_3CH_2CHCH_2CH_2CH_3$$

a. Is he right?
b. How could you redraw the condensed formula to better illustrate the name?

Concept Target

• Practice hydrocarbon nomenclature.

Solution

a. He is incorrect because he didn't use the longest chain of carbon atoms for the root name (it is seven, not six). The name should be 4-ethylheptane.

b. Writing the longest carbon atom chain on a line sometimes is helpful when naming compounds:

$$CH_2CH_3$$
$$|$$
$$CH_3CH_2CHCH_2CH_2CH_2CH_3$$

Conceptual Problem 24.17

Explain why you wouldn't expect to find a compound with the formula CH_5.

Concept Target

• Underscore that in organic compounds, carbon obeys the octet rule (has four bonds).

Solution

Since carbon would have more than four bonds in this case, CH_5 would be in violation of the octet rule.

Conceptual Problem 24.18

Catalytic cracking is an industrial process used to convert high-molecular-weight hydrocarbons to low-molecular-weight hydrocarbons. A petroleum company has a huge supply of heating oil stored as straight chain $C_{17}H_{36}$ and demand has picked up for shorter chain hydrocarbons to be used in formulating gasoline. The company uses catalytic cracking to create the shorter chains necessary for gasoline. If they produce two molecules in the cracking, and 1-octene is one of them, what is the formula of the other molecule produced? As part of your answer, draw the condensed structural formula of the 1-octene.

Concept Target

• Use the description of an industrial refining process to predict hydrocarbon products.

Solution

$CH_2CHCH_2CH_2CH_2CH_2CH_2CH_3$
1-octene

The other product formula is C_9H_{20}.

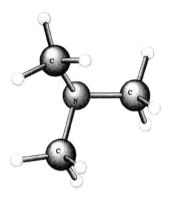

Conceptual Problem 24.19

Given the models shown here where C, N, and O atoms are labeled and H atoms are the small white balls:

a. Write the molecular formula of each molecule.
b. Write the condensed structural formula for each molecule.
c. Identify the functional group for each molecule.

Concept Target

• Interpret molecular models of organic compounds that contain elements in addition to hydrogen and carbon.

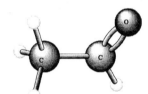

Solution

a. The molecular formulas are: trimethylamine – C_3H_9N, acetaldehyde – C_2H_4O, 2-propanol – C_3H_8O, and acetic acid – $C_2H_4O_2$.

b. The condensed structural formulas are:

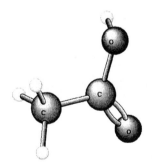

trimethylamine:	acetaldehyde:
	O ‖ CH_3CH
$N(CH_3)_3$	
2-propanol: OH | CH_3CHCH_3	acetic acid: O ‖ CH_3COH

c. The functional groups in the molecules are trimethylamine – amine (tertiary), acetaldehyde – aldehyde, 2-propanol – alcohol, and acetic acid – carboxylic acid.

Conceptual Problem 24.20

Given the models shown where C atoms are labeled and H atoms are the small white balls:
a.   Write the molecular formula of each molecule.
b.   Write the condensed structural formula for each molecule.
c.   Give the IUPAC name of each molecule.

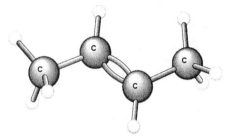

Concept Target

• Interpret complex hydrocarbon molecular models.

Solution

a.   The molecular formulas are *trans*-2-butene –
     C_4H_8, cyclohexane – C_6H_{12}, and
     2,2,4-trimethylpentane – C_8H_{18}.

b.   The condensed structural formulas are:

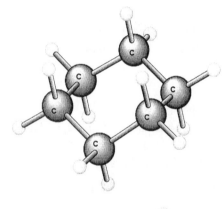

   trans-2-butene:
   $CH_3CH=CHCH_3$

   cyclohexane:

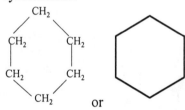

   2,2,4-trimethyl pentane:
   $$\begin{array}{ccc} & CH_3 & CH_3 \\ & | & | \\ CH_3 & CCH_2 & CHCH_3 \\ & | & \\ & CH_3 & \end{array}$$

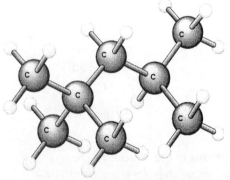

c.   The IUPAC names are *trans*-2-butene,
     cyclohexane, and 2,2,4-trimethylpentane.

Conceptual Problem 24.21

Why would you expect the melting point of the alkanes to increase in the series methane, ethane, propane, and so on?

Concept Target

• Explore the relationship between hydrocarbon molecular weight and intermolecular forces.

Solution

The molecules increase regularly in molecular weight. Therefore, you expect their intermolecular forces and thus their melting points to increase.

Conceptual Problem 24.22

Consider the following formulas of two esters:

$$\underset{\substack{\|\\ CH_3CH_2\text{-}C\text{-}O\text{-}CH_3}}{O} \qquad \underset{\substack{\|\\ CH_3CH_2\text{-}O\text{-}C\text{-}CH_3}}{O}$$

One of these is ethyl ethanoate (ethyl acetate) and one is methyl propanoate (methyl propionate). Which is which?

Concept Target

• Use nomenclature concepts to correctly determine the names of two "new" molecules.

Solution

Ethyl ethanoate (acetate) is:

$$\underset{\substack{\|\\ CH_3CH_2\text{-}C\text{-}O\text{-}CH_3}}{O}$$

Methyl propanoate is:

$$\underset{\substack{\|\\ CH_3CH_2\text{-}O\text{-}C\text{-}CH_3}}{O}$$

Chapter 25

Biochemistry

Concept Check 25.1

Two common amino acids are

$$
\begin{array}{cc}
\begin{array}{c}
CH_3 \\
\mid \\
H_2N-C-COOH \\
\mid \\
H
\end{array}
&
\begin{array}{c}
H \\
\mid \\
HO-C-CH_3 \\
\mid \\
H_2N-C-COOH \\
\mid \\
H
\end{array}
\end{array}
$$

alanine   threonine

Write the structural formulas of all of the dipeptides that they could form with each other.

Concept Target

• Realize that two dipeptides are possible from two amino acids.

Solution

The carboxyl group, —COOH, of either of the amino acids could be bonded through a peptide bond to the amino group, —NH$_2$, of the other. The structures of these two dipeptides are:

$$
\begin{array}{cc}
\begin{array}{c}
\quad\quad\quad\quad H \\
\quad\quad\quad\quad \mid \\
CH_3\; O\; HO-C-CH_3 \\
\mid \quad \parallel \quad\quad \mid \\
H_2N-C-\;\;C-N-C-COOH \\
\mid \quad\quad\quad\quad \mid\;\; \mid \\
H \quad\quad\quad\; H\; H
\end{array}
&
\begin{array}{c}
\quad\quad H \\
\quad\quad \mid \\
HO-C-CH_3\; O \quad\quad CH_3 \\
\quad\quad \mid \quad\quad \parallel \quad\quad \mid \\
H_2N-C\text{———}C-N-C-COOH \\
\quad\quad \mid \quad\quad\quad\quad \mid\;\; \mid \\
\quad\quad H \quad\quad\quad\; H\; H
\end{array}
\end{array}
$$

Concept Check 25.2

Noting the three complementary base pairs and which bases are found in DNA or RNA, write the RNA sequence complementary to the following sequence:

ATGCTACGGATTCAA

Concept Target

• Understand the genetic code.

Solution

In RNA, the bases are adenine (A), guanine (G), cytocine (C), and uracil (U), but not thymine (T). The complementary bases are A and T, A and U, and G and C. So, A in DNA would have U as a complement in RNA; T in DNA would have A in RNA as a complement; G in DNA would have C in RNA as a complement; and C in DNA would have G in RNA as a complement. Here, then, is the RNA sequence complementary to the DNA sequence given in the problem statement:

UACGAUGCCUAAGUU

Conceptual Problem 25.17

It is your job to manufacture polymers from a series of monomer units. These monomer units are called A, B, and C. In this problem you need to "build" polymers by linking the monomer units. Represent the polymer linkages using dashes. For example, -A-B-C- represents a polymer unit made from linking monomer units A, B, and C.
a. Build two different homopolymers from your monomer units.
b. Build a copolymer from A and B.
c. Build a copolymer that is about 33% C and about 66% A.

Concept Target

• Highlight the basic structures of polymers.

Solution

a. Since a homopolymer consists of the same monomer units linked together, examples are -A-A-, -B-B-, or -C-C-.
b. Since a copolymer contains different monomer units linked together, examples include -A-B-, -B-C-.
c. A ratio of 1 C unit to every 2 A units would produce the desired copolymer: -A-C-A-.

Conceptual Problem 25.18

Use resonance formulas to explain why polyacetylene has delocalized molecular orbitals extending over the length of the molecule, whereas the following molecule does not.

Concept Target

- Understand the type of structure that allows for electron delocalization throughout an organic compound.

Solution

Polyacetylene has delocalized molecular orbitals extending over the length of the molecule because double bonds alternate with single bonds over the length of the molecule. Each carbon atom in the chain has one single bond and one double bond to an adjacent carbon atom. The resonance structures are

For the molecule in the problem, the pattern of double bonds and single bonds does not alternate regularly over the entire length of the molecule. Some of the carbon atoms do not have a double bond, and there is no delocalization of electrons in these regions.

Conceptual Problem 25.19

A common amino acid in the body is ornithine. It is involved in the excretion of excess nitrogen into the urine. The structural formula of ornithine is

$$H_2NCH_2CH_2CH_2 - \underset{\underset{NH_2}{|}}{\overset{\overset{H}{|}}{C}} - COOH$$

Write the fully ionized form of the molecule. (Note that there are two ionizable amino groups.)

Concept Target

• Write the zwitterion or fully ionized form given the nonionized form of an amino acid.

Solution

The fully ionized form of ornithine is

$$^+H_3NCH_2CH_2CH_2 - \underset{\underset{NH_3^+}{|}}{\overset{\overset{H}{|}}{C}} - COO^-$$

Conceptual Problem 25.20

Write the structural formulas of all possible tripeptides with the composition of two glycines and one serine. (See the structural formulas in Table 25.1.)

Concept Target

• Write the possible peptides given the amino acids.

Solution

The possible tripeptides with two glycines and one serine are gly–gly–ser, gly–ser–gly, and ser–gly–gly. Here are the corresponding structural formulas:

$$
\underset{\text{gly-gly-ser}}{\underset{\underset{\text{CH}_2\text{OH}}{|}}{\text{H}_2\text{NCH}_2\overset{\overset{\text{O}}{\|}}{\text{C}}\text{—NHCH}_2\overset{\overset{\text{O}}{\|}}{\text{C}}\text{—NHCHCOOH}}}
\qquad\qquad
\underset{\text{gly-ser-gly}}{\underset{\underset{\text{CH}_2\text{OH}}{|}}{\text{H}_2\text{NCH}_2\overset{\overset{\text{O}}{\|}}{\text{C}}\text{—NHCHC}\text{—NHCH}_2\text{COOH}}}
$$

$$
\underset{\text{ser-gly-gly}}{\underset{\underset{\text{CH}_2\text{OH}}{|}}{\text{H}_2\text{NCHC}\text{—NHCH}_2\text{C}\text{—NHCH}_2\text{COOH}}}
$$

Conceptual Problem 25.21

What amino-acid sequence would result if the following messenger RNA sequence were translated from left to right?

AGAGUCCGAGACUUGACGUGA

Concept Target

• Understand the genetic code.

Solution

Mark off the message into triplets, beginning at the left. Then, refer to Table 25.3 to determine which amino acid is represented by each triplet.

AGA | GUC | CGA | GAC | UUG | ACG | UGA

The corresponding amino-acid sequence is Arg-Val-Arg-Asp-Leu-Thr. The triplet UGA is the code to end the sequence.

Conceptual Problem 25.22

Give one of the nucleotide sequences that would translate to the peptide

lys–pro–ala–phe–trp–glu–his–gly.

Concept Target

• Understand the genetic code.

Solution

Refer to Table 25.3 for the sequences that translate to the different amino-acids.

    lys–pro–ala–phe–trp–glu–his–gly

One possible nucleotide sequence is AAA CCU GCU UUU UGG GAA CAU GGU UAA.
The triplet UAA indicates the end of the sequence.